Lecture Notes in Computer Science 1305

Edited by G. Goos, J. Hartmanis and J. van Leeuwen

Advisory Board: W. Brauer   D. Gries   J. Stoer

Springer
Berlin
Heidelberg
New York
Barcelona
Budapest
Hong Kong
London
Milan
Paris
Santa Clara
Singapore
Tokyo

David Corne   Jonathan L. Shapiro (Eds.)

Evolutionary Computing

AISB International Workshop
Manchester, UK, April 7-8, 1997
Selected Papers

 Springer

Series Editors

Gerhard Goos, Karlsruhe University, Germany

Juris Hartmanis, Cornell University, NY, USA

Jan van Leeuwen, Utrecht University, The Netherlands

Volume Editors

David Corne
University of Reading, Department of Computer Science
Parallel Emergent and Distributed Architecture Lab.
RG6 6AY Reading, UK
E-mail: d.w.corne@reading.ac.uk

Jonathan L. Shapiro
University of Manchester, Department of Computer Science
M13 9PL Manchester, UK
E-mail: jls@cs.man.ac.uk

Cataloging-in-Publication data applied for

Die Deutsche Bibliothek - CIP-Einheitsaufnahme

Evolutionary computing : AISB international workshop ... ; selected
papers. - Berlin ; Heidelberg ; New York ; Barcelona ; Budapest ;
Hong Kong ; London ; Milan ; Paris ; Santa Clara ; Singapore ;
Tokyo : Springer
1997. Manchester, U.K., April 7 - 8, 1997. - 1997
(Lecture notes in computer science ; 1305)
ISBN 3-540-63476-2

CR Subject Classification (1991): F.1, F.2.2, I.2.6, I.2.8-9, J.3, J.1

ISSN 0302-9743
ISBN 3-540-63476-2 Springer-Verlag Berlin Heidelberg New York

© Springer-Verlag Berlin Heidelberg 1997
Printed in Germany

Typesetting: Camera-ready by author
SPIN 10545905   06/3142 – 5 4 3 2 1 0   Printed on acid-free paper

Preface

This volume contains selected papers from a workshop on Evolutionary Computing which took place in Manchester, April 7–8, 1997, as part of the Society for the Study of Artificial Intelligence and Simulation of Behaviour (AISB) workshop series. This workshop was open to submissions from all aspects of evolutionary computing, such as genetic algorithms, genetic programming, and other population-based, adaptive computing systems. The workshop consisted of thirty papers reporting on a range of topics, including: mathematical and experimental studies of the performance of evolutionary algorithms, novel and high-performance applications of evolutionary computation, and computational models of evolutionary systems, including biological systems and economic systems.

After the workshop, the authors were asked to revise their papers and resubmit them. These were then sifted through a further refereeing and revision process. Those submitted papers which the referees and editors felt were of the highest quality are presented here. In selecting papers for inclusion, we have made no attempt to ensure wide coverage by topic or geography. Nor have we treated posters and oral presentations differently. We simply asked the referees to identify the papers which were the most worthy of publication, and to give feedback to the authors to improve quality and clarity. We hope that this has resulted in a collection of high quality and high readability, and that researchers and students of evolutionary computing will find this volume stimulating and useful.

The AISB workshop on Evolutionary Computing was started by Terry Fogarty in 1994 as a forum in which UK-based practitioners could meet and discuss their research. It has since grown so that this most recent workshop was considerably more international. The intent has not changed, however, which is to provide a technical, yet informal small workshop, which would be easily affordable to attend and which would allow for discussion and interaction as well as the presentation of polished work.

In addition to the submitted papers, there were two invited speakers at the workshop. Stan Metcalfe of Manchester University spoke on "Fisher's Principle in Evolutionary Economics". In this presentation, he reviewed how evolutionary and population-based dynamics are being used to model and understand the changes in economic entities, such as firms and businesses. Terry Fogarty gave a presentation on "EVONET: Current Activities and Future Plans". EVONET is a so-called Network of Excellence funded by the European Union to coordinate research, training, and technology transfer among the key academic and industrial researchers in evolutionary computing. Unfortunately, due to other commitments, neither speaker was able to contribute to this volume. In order to give the readers some access to this part of the workshop, we provide some starting points. Readers who want to follow up on the work that Professor

Metcalfe presented could look at his 1994 article in the Journal of Evolutionary Economics (volume 4, pages 327–346), or at his forthcoming book: "The Evolutionary Economics of Creative Destruction: The Graz Schumpeter Lectures". Readers can find out more about EVONET at the EVONET website (http://www.dcs.napier.ac.uk/evonet/evonet.htm).

There was a panel discussion at the workshop, chaired by Riccardo Poli, in which participants discussed their experiences working in industry and with industrial partners. The goal of this discussion was to find how to best facilitate interactions between academic and industrial workers. One of the organizers (D.C.) had the bright idea to hold this event in a local public house. This had the happy consequence that many workshop participants felt compelled to show their gratitude to the organizers by buying them drinks. This had the less fortunate consequence that the other organizer, the one writing this (J.S.), cannot remember exactly what was concluded by the panel (I do seem to recall that the sun was shining). One point which was stated emphatically by several participants was that academic researchers should insure that they are involved with the people within a company who "own the problem" rather than with that unit of the company whose job it is to solve such problems.

It is a pleasure to thank several people and organizations for help in producing the workshop. First, we thank the substantial program committee who did most of the first round of paper refereeing (to sanction presentation at the conference). Next, we thank the additional referees who joined the program committee to help us in meeting tight deadlines for the second round of reviewing. The entire group are listed on the next page; their efforts are sincerely and greatly appreciated. Additional refereeing was done by the organizers.

The AISB, particularly Medeni Fordham, and the Department of Computer Science at Manchester University, particularly Angela Linton and David Bree, provided a large amount of organizational support for the workshop. We also thank the EPSRC for providing funds which allowed postgraduate students to participate, and the Royal Society for providing the funds which allowed Dr. Kalmykov to travel from Russia. We also thank Terry Fogarty, Jan van Leeuwen and Alfred Hofmann, who each provided valuable support and assistance in the preparation and production of this volume.

Finally, it was gratifying that so many good researchers chose to participate in and contribute to this workshop. Ultimately, only the content of the workshop really matters. If it was good, it was the participants who made it so. We greatly appreciate their contributions. We hope this volume will help inspire further research into this exciting field.

July 1997   David Corne
                                                                         Jon Shapiro

Organization

The AISB Workshop on Evolutionary Computing 1997 is one of several workshops sponsored and organized by the AISB on an annual basis. In 1997 the workshops were hosted by the University of Manchester.

Organizing Committee

General Co-Chair:	David Corne (University of Reading)
General Co-Chair:	Jon Shapiro (University of Manchester)
Panel Chair:	Riccardo Poli (University of Birmingham)

Referees

Edmund Burke,	University of Nottingham
Hugh Cartwright,	University of Oxford
Marco Dorigo,	Free University of Brussels
Bruce Edmonds,	Manchester Metropolitan University
Stuart Flockton,	London University
Terence Fogarty,	Napier University
Pete Hancock,	University of Stirling
Inman Harvey,	University of Sussex
Robert Heckendorn,	Colorado State University
Phil Husbands,	University of Sussex
Dirk Mattfeld,	University of Bremen
Ben Paechter,	Napier University
Ian Parmee,	Plymouth University
Ray Paton,	University of Liverpool
Riccardo Poli,	University of Birmingham
Nick Radcliffe,	Quadstone Ltd, Edinburgh
Soraya Rana,	Colorado State University
Magnus Rattray,	Aston University
Colin Reeves,	Coventry University
Peter Ross,	University of Edinburgh
Andrew Tuson,	University of Edinburgh
Darrell Whitley,	Colorado State University

Sponsoring Institution

The Society for the Study of Artificial Intelligence and Simulation of Behaviour (AISB), University of Sussex, Brighton, BN1 9QH.

Table of Contents

Evolutionary Approaches to Issues
in Biology and Economics

Simulating Pricing Behaviours Using a Genetic Algorithm

Sue Bedingfield, Stephen Huxford, Yen Cheung,
Dept. of Business Systems, Monash University, Clayton, 3168, Australia

Abstract
Retail petrol prices in Australia are monitored by the federal government which sets the base price for petrol that oil companies must follow. Even though current regulations prohibit the companies from colluding, some flexibility over the actual retail price of petrol is allowed. This paper examines the oligopolistic behaviour of the petrol sellers in the API (Australian Petroleum Industry) using game theory and a genetic algorithm (GA). Experiments based on the API retail marketplace interaction were conducted with particular consideration given to the API rebate system. The major oil companies may set their petrol price below a fixed target price, but if they do so, they must rebate their sellers with the difference. Initial results suggest that game theory concepts and GA's are suitable tools for studying the API. Further work related to this project includes incorporating more realistic constraints into the system, better representation of the data in the model as well as comparing the results with human experts.

Keywords: genetic algorithm, Iterated Prisoner's Dilemma, petrol pricing

1. Introduction

The API can be regarded as an oligopoly because it includes only a relatively small number of companies and the pricing behaviour of any particular company will influence the decision making and pricing strategies of other participants in that industry. In this project, the decision making behaviours of wholesale sellers in the API were modelled using game theory concepts. The initial model comprises four wholesale petrol sellers (as there are four major oil companies in the API) . The main pricing criteria is based on the simple law of economics where the market share is increased when the price of petrol is decreased. The API's rebate system is incorporated in the model to provide some realistic features of the industry.

Traditional economic models, equation-based models and expert systems cannot cope easily with the complexity of a dynamic marketplace such as the API. Due to the computational demands of evaluating the extensive set of possibilities these approaches are not as efficient as the evolutionary approach. In the evolutionary approach, the companies are modeled as players with specific (pricing) strategies which enable them to interact dynamically with each other during the simulation. This interaction is easily modeled using basic GA concepts. Background literature on GA's can be found in [1],[2],[3],[4]. The following sections describe the approach and the results obtained from the simulations.

2. The Prisoner's Dilemma Game

First discussed by A W Tucker in 1950, the Prisoner's Dilemma (PD) game was used to demonstrate the difficulty of analysing certain games. Further information on the PD game can be found in [5]and [6]. For modelling business interactions such as the API, the N-person Iterated PD game is more relevant to business applications. For example, when determining the price of petrol, the pricing manager takes into account the number and nature of other competitors in the market. Also, the interactions are expected to be over a long period of time as the companies would wish to remain in the marketplace as long as possible. The N-person IPD behaves differently from the one shot PD game in that in the former case the dominant strategy is to '*co-operate*' whereas in the latter case, it is better to 'defect' (see Table 1 below). There is much research on the evolution of co-operation amongst a number of players using evolutionary computing ([6], [7],[8]).

In general, the payoff matrix for the game (see Table 1) is defined according to the following constraints ([9]):

$$2a < b + c$$
$$c < a < d < b$$

where   a is the payoff for mutual co-operation (i.e. 3);
           b is the payoff for co-operating when other player defects (i.e. 0);
           c is the payoff for defecting when other player co-operates (i.e. 5);
           d is the payoff for mutual co-operation (i.e. 1).

From Table 1, it can be seen that even though the dominant strategy is to defect, the players will be worse off if both of them decide to defect.

Case	P1	P2	Score (P1)	Score (P2)
1 (CC)	Co-operate	Co-operate	a	a
2 (CD)	Co-operate	Defect	b	c
3 (DC)	Defect	Co-operate	c	b
4 (DD)	Defect	Defect	d	d

Table 1: Scoring System of the PD Game

3. Problem Description

Accounting for around 40% of the total domestic energy supplies in Australia, the API comprises three sectors: refining, wholesaling and retailing. In order to protect consumers, companies are not allowed to collude in their pricing strategy. In practice collusion may often 'appear' to be the case as seen in the thirty inquiries into the industry's pricing strategy during the last two decades ([10]).

The model developed in this project is based on the legal requirements that there is no 'communication' between the sellers (as in the PD game) when they are determining their pricing strategy. A rebate system is employed to compensate the petrol retailers when their supplier (major oil company) decreases the retail price of the petrol. This may occur when a wholesaler reduces the retail price in order to increase market share and is usually a short term decision. Perturbations reflecting this behaviour have been observed in the model. Experiments were conducted for a fixed total market size, i.e. the demand for the petrol is constant independent of the average retail price.

3.1 Implementation Details

In the initial model, consideration was given to factors such as retailing margins, the number of competitors, historical information and the size of the required rebate. Other relevant factors such as consumer behaviour and oil distributions costs will need to be incorporated at a later stage.

To begin with, it is assumed that profit is a linear combination of retail price (and rebate). In order to make the problem combinatorialy tractable a number of assumptions were made. It is assumed that at any point in time each player knows their own price and their own ranking with respect to the price of the other players for the previous 2 rounds. However the actual identity of the other players at each particular rank is not known. We assume that there is only one price per competitor. If n is the number of competitors, at any time t, there are n_t distinct prices, resulting in n_t possible rankings for any particular competitor.

i.e.

n = number of companies/players involved

At any time t, there are n_t different prices where $1 \leq n_t \leq n$.

Since n_t can be any value from 1 to n, the total number of possible rankings is therefore:

$$\sum_{k=1}^{n} {}^{n-1}C_{k-1} \times k \qquad (1)$$

and the possible number of 2-round histories is:

$$\left\{ \sum_{k=1}^{n} {}^{n-1}C_{k-1} \times k \right\}^2 \qquad (2)$$

For example, a possible 2-round history for a particular player in a 4-player game could take the form :

history 1	2: 1,1,2,3
history 2	3:1,2,3,4

where history 1 is interpreted as the player having the second highest price and there were 2 players with the highest price, and one for each of the other two prices. Similarly for history 2. So for a 4-player 2-history game the number of possible ranks is 20. In order to reduce the complexity of the problem, we defined 3 possible actions as shown below. Hence, given any particular history a player can make one of the following 3 moves:

- increase the price by one unit
- do not change the price
- decrease the price by one unit.

The model was implemented in Object Pascal in the Delphi Environment. This rapid application development tool includes features for database manipulation, reporting and user interface development.

3.2 Representation of Marketplace Participants

The representation of players in a marketplace is an extension of the scheme used by Axelrod ([11]) to represent players in an IPD game. Each individual's chromosome embodies its complete strategy in the marketplace, i.e. it enumerates the individual's response to every possible ranking history. Experiments were conducted using four-player[1] competitions where each player can remember the ranking configurations of the two most recent competition steps. Since the possible response to each ranking configuration is a 3-way choice, a 3-nary string of length 400 was used to contain a player's complete response repertoire. Each component of a string corresponds to a specific ranking history. The player's response to this particular history is contained within the component.

---

[1] This number is close to the actual number of participants in the API.

[2] 52 was used in these experiments, but can be varied

3.3 Playing the Game

The simulation begins by selecting a population size (*popsize*) of 20 players. In order to calculate the fitness, the simulation proceeds with the players using their strategy (byte string) to play against each other. The algorithm proceeds in the following way:

- Select all possible combination of 4 players (chromosomes) from the population (with replacement). There are $^{20}C_4$ of these combinations. Each player sets their starting price at a predetermined base level.
- Play these 4 players (chromosomes) against each other for 52 iterations[2] as follows:
  - Based on the 2-move history each player makes a move,
  - The resulting prices are recorded in a table and the ranking of each player is incorporated into an updated 2-move history,
  - The profit for each player is recorded,
  - The ranking of the players is recalculated using the new prices and recorded.

- The chromosomes are then ranked by total accumulated fitness and the process of selecting a new population is performed using the roulette wheel approach ([2]).

3.4 Genetic operators

Whether to mutate or not is decided in the usual way. If mutation is needed then the decision that the chromosome makes for that particular history is randomly changed. For the 3-nary string representation described, one point crossover was adopted. This makes sense for the representation described and does not require any modifications such as a repair operator.

3.5 Fitness Criteria and Market Share

As mentioned, we include margins, number of competitors, histories and rebate details in our representation. The fitness function per time step becomes:

$$\left[(p_i - C) - \delta_i(R - p_i)\right]VS_i$$

where p_i = the price charged by seller i

C = cost price of the petrol

V = total volume of sales in the current time step[3]

R = rebate price

$$\delta_i = \begin{cases} 1, p_i < R \\ 0, p_i \geq R \end{cases}$$

S_i = is seller i's share of the total market (so $\sum_i S_i = 1$)

---

[3] Both cost and volume were assumed to be constant for this particular experiment

Since the market share of an individual is in inverse proportion to the cost, we define S_i as follows:

$$S_i = \frac{\sum_{j \neq i} p_j}{3 \sum_{j=1}^{n} p_j}$$

4. Analysis and Discussion of Results

Results of this simulation are displayed in Figures 1 and 2. Figure 1 displays the fitness of the fittest individual per generation and the average fitness per generation. As seen from Figure 1 the fitness produced by random player strategies improved significantly throughout successive generations compared with the initial fitness, and has stabilised by 100 generations.

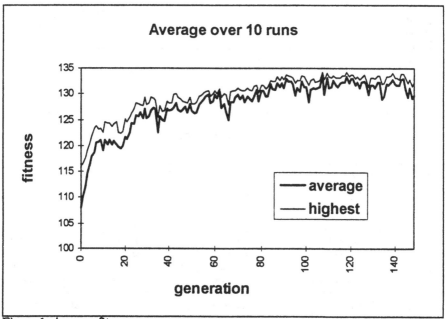

Figure 1. Average fitness

Figure 2 demonstrates the percentage increase in price per generation for both the highest and lowest increases. This percentage is calculated by comparing the actual amount of the overall increase with the greatest possible amount of increase, i.e. (1 price unit) $\times$ (number of competition steps). It can be seen from the graph that initially the tendency is to increase the price but by less than 50% of the maximum possible, in the case of the players with the greatest increases. In the case of the players with the lowest increases, the overall tendency initially is to decrease prices by up to 20% of the maximum possible decrease. However, ultimately these initial tendencies disappear and the predominant behaviour is to increase prices. The reason for this is that if a player decreases their price relative to the prices of the other players, the effect is to increase their market share, however not enough to compensate for the reduction in profit. The more successful players, i.e. the fittest and therefore those that survived, had a strong tendency to increase their prices. We suspect that some players have effectively learnt to follow a leader in the marketplace which is what actually happens in the API.

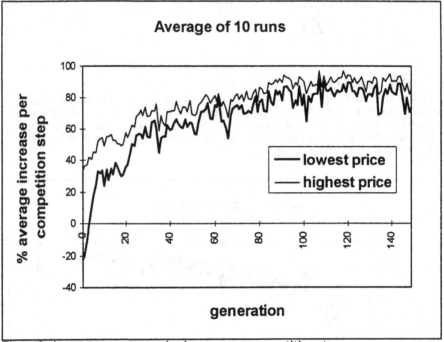

Figure 2. Average percentage price increase per competition step.

Experiments involving a variable change of total market size have been studied and reported in [12]. This was based on the assumption that if the average price of a product increases, the market demand may decrease accordingly. However, the demand in the API remains almost constant as the average price varies. Modelling the variation in total market size may be more appropriate to other types of industries where demand is more sensitive to price. The results for both the fixed and variable markets were similar however, in the variable market case, the equilibrium level is not as high.

5. Further work
A problem with the current implementation is the use of byte strings. As the number of players, remembered histories and possible number of choices or moves are increased, the memory requirements can be unmanageable. Current investigations on this project also involve incorporating further price determinants into the fitness function. Investigations on the assumed market share ideas in this paper are also being pursued. Other related work includes further analysis, identification and interpretation of a good strategy for the API. Also, collection of real data is being conducted in order to compare the performance of the genetic algorithm with actual performance.

6. Conclusion

Initial results suggest that the API can be modelled using game theory concepts and that GA's are suitable tools for coping with the large number of competition steps in the marketplace when determining the characteristics of a good strategy for the API. However extensions to the basic concepts of both game theory and GA's are necessary in order to conduct a more rigorous study of the API.

References
1.   Goldberg, D(1989)*GAs in Search, Optimisation and Machine Learning*, Reading, MA: Addison-Wesley.
2.   Michalewicz Z (1993), *Genetic Algorithms + Data Structures = Evolution Programs*, Second, Extended Edition, Springer-Verlag.
3.   Holland J (1975) *Adaptation in Natural and Artificial Systems*, Ann Arbor:University of Michigan Press. (Also Second Edition, MIT Press, 1992).
4.   De Jong, K (1975), *The Analysis of the Behaviour of a Class of Genetic Adaptive Systems*, PhD diss., University of Michigan, Ann Arbor December 1991.
5.   Tucker A (1980) *A Two Person Dilemma*, UMAP Journal, pp 101-1-3.
6.   Axelrod R(1980)*Effective Choice in the Prisoner's Dilemma* Journal of Conflict Resolution, pp 3-25, March '80.
7.   Fogel D(1993) *Evolving behaviours in the iterated prisoner's dilemma*, Evolutionary Computation, 1(1):77-97.
8.   Yao, X, Darwen P J (1994), An Experimental Study of N-Person Iterated Prisoner's Dilemma,   Informatica 18, pp 435-450.
9.   Fogel D (1995) *Evolutionary Computation*, IEEE Press.
10. 'The Right Price? Profitability, Investment, Competition and The Consumer: A Report on the Australian Petroleum Industry', ACCESS Economics, pp82.
11. Axelrod R (1987) *The evolution of strategies in the iterated prisoner's dilemma*, In L. Davis, ed. Genetic Algorithms and Simulated Annealing, pp 32 - 41, Morgan Kaufmann.
12. Yen Cheung, Sue Bedingfield, Stephen Huxford, 'Oligopolistic Behaviour of the Australian Petroleum Industry', Proceedings of International Conference on Computational Intelligence and Multimedia Applications, 10-12 Feb, 1997, Gold Coast Australia, Ed. B Verma & X Yao, pp 161-165

Biologically Inspired Computational Ecologies: A Case Study

Paul Devine and Ray Paton

Department of Computer Science
University of Liverpool
PO Box 147
Liverpool L69 3BX, UK
email: paul@csc.liv.ac.uk
tel : 44 0151 794 3692

Abstract. Some aspects of evolution are, by their very nature, unsuited to a process of direct experimentation. The work described here is a computational system strongly inspired by real ecology, it is intended as a framework within which the interaction of evolution, learning and cultural effects may be investigated. The design, development and behaviour of the system is outlined in some detail.

1 Introduction

"Augescunt aliae gentes, aliae minuutus. Inque brevis spatio mutantus saecla animantum. Et quasi cursores vitae lampada tradunt" **Lucretius** *– De Rerum Naturae ii.7*

The computational ecology described here was conceived as a system for the investigation of a number of aspects of natural adaptation and their interrelationships, this is an extensive development of earlier work [DKP96]. The implementation is a multi-agent system, the agents being inspired by the animat approach [Wil90] and their environment derived from a number of salient ecological principles. Special attention has been paid to developing this environment in respect of resources and defining agent-agent and agent-environment interactions in the context of their biological equivalents. Briefly, a herbivore ecology is simulated, the mechanisms of adaptation being centred around a classifier-type system. It must be emphasised that the systems in question here have been developed from problems and models located in the domain of ecology. The principal aim of the approach utilised is to parallel apsects of real ecologies in an artificial system. This may then be used to formulate and explore hypotheses such as the relationships between different levels of adaptation, cultural effects and ratios of sexual and asexual reproduction in hermaphroditic populations.

2 Motivation

The motivation behind this work was manifold, the construction of a biologically inspired artificial ecology being of interest in itself. In common with other workers we were attracted to simulated evolution [DP97] as a means of addressing problems unsuited to other forms of experimentation due to the processes and timescales involved in their natural equivalents. This would enable us to set and test hypotheses which are sensitive to questions posed by behavioural ecologists. In our case we have worked with Prof. Geoff Parker of the Dept of Environmental and Evolutionary Biology, University of Liverpool. Also, the agent based approach permits the construction of individual based models [Jud94]. These are useful for a number of reasons, two in particular stand out. Firstly, evolution is fundamentally selectionist and hinges on the fact that individuals in a population are non-identical. Secondly, even discounting evolutionary considerations individual variation has other effects on populations and their demography [Lom78]. Furthermore, the effects of learning on evolution were of particular interest. This issue has been of interest for over a century [Bal96] and a lot of recent work in has revolved around this and related topics, for example [AL91], [FM94], [PNC92]. One of our interests is in how simple cultural learning systems may affect the evolutionary process. Here horizontal, diagonal and vertical transmission are all of importance rather than just parent-progeny transmission, for example by imitative learning [CMB95]. A simplified representation of the system is illustrated in Figure 1, this shows a single agent and its immediate environment, the ovals represent other agents. Much of the detail is explained later, this figure is intended to provide a frame of reference for subsequent discussion.

3 Relationship to Similar Systems

Given the underlying motivation of using a computational system to address genuine ecological research problems we have developed a simulation system called 'Herby', a plant-herbivore simulation system. This in part overlaps a number of other systems but differs from them all in either architecture, motivation, implementation or all three. For example, some of the other work addressing the relationships between learning and evolution includes Latent Energy Environment based work [AL91], this is fundamentally different in its methodology. The work of French and Messinger [FM94] is more highly abstracted and problem specific.

The two extant systems which most closely resemble that described here are Swarm [Hei94] and Echo [Hol92], [FJ94]. These are in many respects larger, more complex undertakings. Swarm is essentially a framework within which a large variety of complex adaptive systems may be modelled, ecologies being just a subset of these. Echo is probably the closest in conception, but nevertheless we believe it to be quite distinct. Like Herby, Echo is spatially explicit and agent based, however its treatment of agents, resources and interactions is significantly different. Echo agents interact by combat, trade and sexual reproduction. In our system interactions are limited to breeding, the implicit trophic interactions of an interferential grazing system [Mon67] and the exchange of information between geographically connected agents.

The Echo approach allows a wider range of interactions to be modelled in an abstract manner, ours is more specific to a subset of ecological systems. Echo agents migrate by default, movement in Herby agents is 'intentional', as is direction. Resources in Echo are also more abstract than those in Herby, the latter being designed to replicate many facets of a specific type of resource rather than a more general concept.

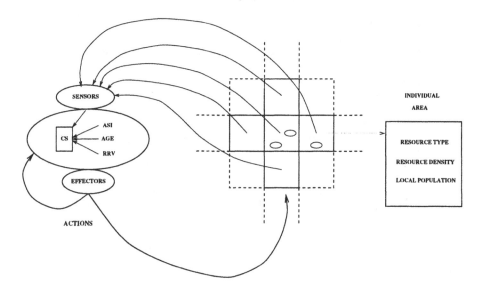

Fig. 1. A single agent and its environment

4 Adaptation and Variation: The Modified Classifier System

In order to investigate adaptation in an evolutionary context a mechanism for both the representation of variation and the mechanisms of heredity and adaptation was required. A highly modified classifier system [BGH90] was developed as the adaptive engine for each agent, this is shown in Figure 2. The classifier as conceived here was not intended to perform complex machine learning tasks it has been directed towards in other work, eg. [Wil90], [MG94]. It was intended purely as a mechanism of basic adaptation. This simplification was necessary in order to run simulations containing large numbers of autonomous agents over very long virtual time periods.

The classifier system (CS) employed follows the Michigan style [Hol92] which has been modified in order to suit the type of application considered here. The primary modification has been the replacement of the genetic algorithm used for rule discovery by a two tier selectionist architecture. Each agent has its own set of rules (rule base) together with the other components associated with classifier

	Echo	Herby
Spatial	Explicit	Explicit
Interactions	Combat Trade Breeding	Breeding Scramble Competition Information Exchange
Movement	Automatic	Agent selected behaviour
Mechanism	Tab Based	Classifier Based
Resources	Discrete resources at discrete locations	Continuous resources heterogeneously distributed

Table 1. Some differences between Echo and Herby

systems, i.e. a message board and a mechanism for the operation of a bucket-brigade algorithm.

Individual rules are comprised of condition/action productions made up of the alphabet [0,1,*], each with an associated strength. The condition string of the rule is at present seventeen bits long. Various segments of this rule string may correspond to differing aspects of the environmental and somatic input. Immediate action is determined from the final bits of the action string topping the message board each round.

CSs commonly utilise GA's, these act on the rule population in order to produce fitter rules by the mechanisms of selection and the usual genetic operators. The GA is effectively the source of novelty and rule discovery. Here the strings are the individuals that comprise the population and the role of the GA is to evolve optimal rules. The CS envisaged here was to form both the adaptive mechanism of an individual and the representation of variation, an individuals "identity" was to reside, at least in part, within the rule set. The genetic operations were to apply only upon the reproduction of an individual. Crossover requires the interaction of two individuals in sexual reproduction. Mutation and inversion may apply to both sexual and asexual reproduction. This restriction on the action of genetic operators limits rule discovery to reproduction, new rules only being generated between generations. This leaves an individual with no means of adaptation other than altering the weights of already extant rules. In order

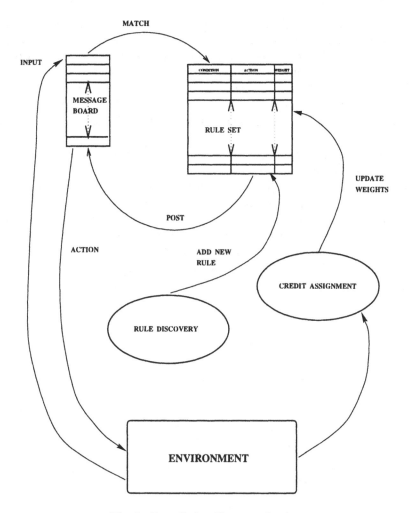

Fig. 2. Overall classifier organisation

to overcome this difficulty there were two options. The first was to implement a second GA at a level that operated during an agent's life-cycle, this would use the initial rule base as the substrate and of necessity make fundamental alterations. This may have been overcome by retaining the original rules and using those alone for reproduction. The second approach was the use of a substring matching algorithm. As the goal was a system that would run comfortably on an HP RISC workstation the second method was chosen.

4.1 Rule discovery mechanism and rule-set partitioning

Rule discovery operates over two distinct timescales, on an evolutionary basis from generation and on a somatic basis within the life-cycles of individuals. It is this latter mechanism which is discussed here. As previously described, the condition part of a rule has areas within it that relate to specific inputs, this may logically be divided into substrings representing various aspects of sensory input, ie. the substrings would correspond to different features of the surrounding environment and an agent's state. The extant rule base may then be searched for matching substrings rather than complete rules and new rules may be recombined from suitable components. This leads to a partition within the rule base in accordance with the hereditary and non-hereditary nature of the respective rules.

The two-tiered rule discovery mechanism is centred on this segregation of S into these two distinct sets, one generated during somatic time (i.e. within an agent's life-cycle) (S") and the other persisting over evolutionary time (S'). An agent is 'born' with S' inherited from its parent, or parents depending on the type of reproduction used. At birth S" is empty but may be gradually populated with new rules generated by the action of the substring matching process described above. S" may also be directly augmented by adding in a complete rule from an external source. When viewed from the agent level this is analogous to direct learning and is important in the formation of the 'cultural' rule set outlined later.

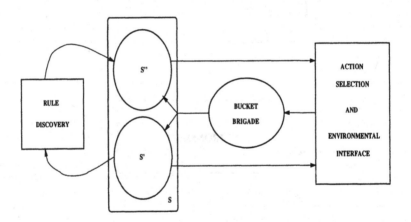

Fig. 3. Role of rule set in classifier operation

There are two mechanisms by which S' may be transmitted to the next generation; sexual and asexual reproduction. Asexual reproduction simply involves the progeny of an agent inheriting S' into its own classifier system with the possibility of the action of mutation; this acts at random points in S'. Sexual reproduction occurs between two agents and involves multipoint crossover on the both sets.

Additionally, we have implemented a form of information exchange that permits rule exchange between classifiers under certain circumstances. As each agent incorporates its own unique classifier system this allows the possibility for rules contained in S" to cross generations as a consequence of cultural processes.

5 Ecological Aspects

Natural ecologies, like many complex systems, possess properties that render them extremely difficult to investigate with the use of traditional formal models that rely heavily on the application of analytic mathematical techniques [CWFL93]. The advantages of the application of adaptive agents to individual based models are twofold. Firstly, they allow the development of ecological models that provide complementary perspectives to extant traditional models as variation in behavioural and physical traits is explicitly represented. Secondly, this explicit representation of variation may permit the investigation of evolutionary processes that are impossible to observe in the field due the long timescales involved.

One of the most important features of natural evolution is the fact that populations are not usually homogeneous but are made up of non-identical individuals. This variation is the essence of selectionist systems which have been described by Darden and Cain [DC89].

Even over non-evolutionary timescales individual variation may have considerable impact on the dynamics of a system. This has been demonstrated by some individual-based field work. The following examples [HDP84] provide straight-forward illustrations of analysis at this level of the population. The development of size distribution in plant populations is very sensitive to the individual organisms involved and slight differences in initial conditions. Light is an essential resource for plants and they compete for it; consequently relative differences in plant height are critical to the outcome of this competition. If there is a low initial variation in height then the final population will be relatively homogeneous in size. Conversely, a high initial variation confers a competitive edge on the taller individuals which will capture an increasingly large portion of the available light and as a result grow at a higher rate. This positive feedback will accelerate the divergence of sizes and the final population will consist of a few very large individuals and a number of much smaller ones. Another instance of the importance of slight differences in initial sizes of individuals is a series of experiments involving populations of 250 bass in 50 gallon aquaria. Cannibalism is common in bass and it has been observed that a bass is capable of eating another bass up to half of its own size. The corollary is that in cases where there is a large initial variation the larger bass tended to eat the smaller bass and therefore grew rapidly; a low initial variation led to a larger population of smaller individuals as a consequence of the lack of intra-species predation.

5.1 Resources, their nature and structure

Much conventional ecological theory revolves around resources and competition for those resources. The model proposed here is no different in that resources are central, they are diverse and an appreciation of their nature is essential. Before proceeding further a definition of resource is necessary. Like many commonly used terms there seems to be no universal definition; for the present purposes that provided by Wiens is applied [Wie84].

"An environmental factor that is directly used by an organism and may potentially influence individual fitness."

These factors may take many forms, in the present context we deal exclusively with food as a resource and restrict agent-environment interactions to the purely trophic. Though the rest of this discussion is general in nature it may equally be applied to "food" resources. Resources vary in a number of ways, Price [Pri84] provided the broad temporal categorisation of resources (fig 2) which we briefly summarise below.

- *Rapidly increasing resources* generally do so over a season and then decline rapidly as the season ends.
- *Pulsing resources* both increase and decrease rapidly with long intervening periods of unavailability.
- *Steadily renewed resources* are produced over long periods and not generally over exploited.
- *Constant resources* are usually uninfluenced by factors such as seasonal change or exploitation, space being a particularly good example in many systems.
- *Rapidly decreasing resources* are produced in a short period of time after which they decline.

Spatial aspects of resource availability are important too. Many are not continuous in their distribution over the environment but are "patchy".

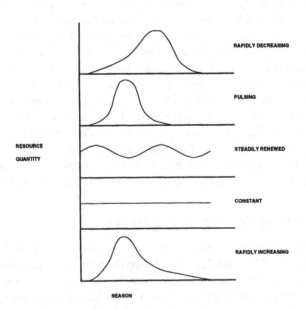

Fig. 4. Resource Types. Adapted from Price

In addition to the nature of resource renewal and its spatial/temporal distributions, the nature of the system components must also be addressed.

5.2 Ecosystem organisation

It goes without saying that the complexities of biological ecosystems are far beyond what it is currently possible to simulate on a computer. Indeed, Levins [Lev66] has noted that no single model may simultaneously optimise for generality, precision and realism. A plant-herbivore system was chosen as the model subject as it lent itself to simulation without too great a degree of simplification. However, even the complex model described here embodies a high degree of abstraction. On the face of it the absence of predation may appear to be an oversimplification. However, it is not unusual for such systems to be little impacted by predation and at the extremes of r-K selection continuum [Pia74] that are most likely to emerge in the simulation predation has very little impact. For example, elephants and lemmings occupy the far ends of this spectrum and predation has little effect on the population dynamics of either species. For the initial work a steadily renewed resource pattern was chosen in order to avoid seasonality and provide the agents with an environment that possessed a simple but realistic pattern of predictability in respect of resource properties.

In terms of the components shown in Figure 5 the resource is flora, its abundance being dictated by various environmental attributes and the translators affecting availability are the herbivorous agents themselves. The behaviour of these agents constitutes resource use and is linked both to translation and physiological processing in that the agent decides to allocate the resource to movement, breeding or its own somatic state. The effects on individual fitness is obvious and the consequent effects on community and population structure formed the basis of the initial investigation.

This herbivore system was developed following the classification scheme of Coughley and Lawton [CL81] which derived its primary division from Monro's dichotomy [Mon67], as shown in Figure 6. In this context 'herbivore' refers not only to cows, sheep etc. but to all creatures feeding on vegetable matter and 'grazing' encompasses all eating of plants by animals.

Non-interactive systems are typified by the herbivore's inability to influence the rate at which its resources are renewed. Further subdivision produces:

- *Reactive* systems where the rate of increase of the herbivores reacts to the rate at which the resources are renewed
- *Non-reactive systems* where it is independent of the of the parameters of resource renewal.

In interactive systems the animals do influence the rate of plant renewal, and this in turn influences the rate of increase of the animals. Further subdivision here gives:

- *Laissez-faire systems* where the herbivores do not interfere with with each other's feeding activities

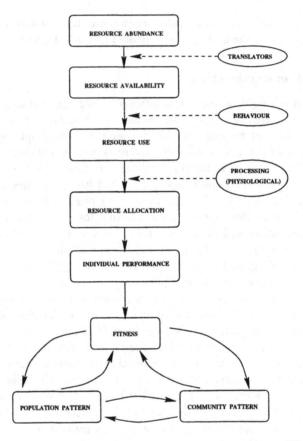

Fig. 5. Components of a Resource System. Adapted from Wiens

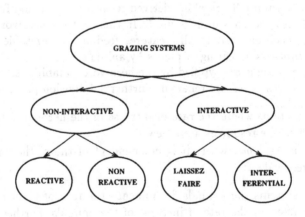

Fig. 6. Herbivore Classification

- *Interferential* systems where the animals do interfere with each other's ability to obtain food.

5.3 The environment

The topology of the environment is toroidal. This arrangement was chosen to eliminate the "edge effects" that would result from the presence of boundaries. These would add considerable problems for the agents to overcome. Their behaviour and evolution was in the context of resource exploitation was under investigation rather than any "maze following" type problems. However, it is easier to conceptualise the environment as a two dimensional grid, here it accords more closely with the agent's relationship to it.

Environmental context is heavily emphasised in the design motivation, the features that are most important to capture are the spatial and temporal arrangement of resources and the agents' interactions with these resources. The resources available were sessile and purely trophic in nature and can be thought of as flora. There are a possible four types of flora present, each with its own nutritional value and growth rate. There is an upper bound on the amount of flora an individual area may support, this is effectively a space constraint.

5.4 The agent

An agent possesses a number of simple attributes that govern its state and is capable of acting in a number of predefined, primitive ways. More complex behaviours emerge from the sequential combination of these behavioural primitives by an agent. Figure 7 shows the interactions between an agent and its environment. Population refers to the entire agent population and community to those it may directly interact with due to geographic distribution. The environment is the entire set of areas open to an agent and a patch those areas it can perceive or directly interact with or move to in a round.

The agent's state variables are borrowed from theoretical ecology [Beg85] but have redefined to suit the application. These are:

Accumulated Somatic Investment (ASI). Here this is related purely to the amount that the agent eats and the energy that it expends.

$$ASI_a(t+1) = ASI_a(t) + f(d, N, e) - \gamma \tag{1}$$

Here a is an individual agent, γ the maintenance cost per round and $f(d, n, e)$, the trophic uptake includes the number of agents eating in the immediate area, N (as the system is interferential), the density of the flora, d, and its nutritional value, e. Should an agent's ASI fall to zero it is killed off. ASI is also reduced by reproducing. In some simulations it was found to be useful to place an upper bound on ASI as shall be discussed later. Residual Reproductive Value (RRV) is a measure of the reproductive success of an agent, it has been implemented in two forms, the latter being the most important and closest to ecological theory.

$$RRV_a(t+1) = RRV_a(t) + \beta \tag{2}$$

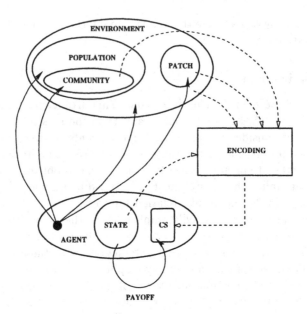

Fig. 7. An Agent's Interaction with its Environment. Broken arrows represent information and solid arrows actions

β is a constant payoff. (2) only applies on rounds when an agent breeds. This has only been implemented in a version where reproduction is exclusively asexual. The second form has been implemented in systems where reproduction is both sexual and asexual.

$$RRV_a(t+1) = RRV_a(t) + \alpha.D \tag{3}$$

Here D is the number of living descendants an agent has and α is a fixed constant. Age is simply the number of rounds that the agent has lived through, a maximum age is usually set for a simulation, after this an agent is automatically killed off. Overall fitness is determined by the simple expedient of summing RRV and ASI. Consequently coefficients and functions are chosen to prevent one dominating the other too much.

Simple life histories may be developed by setting bounds on state information, those used here are a minimum age and ASI at which a reproduction may occur. There is no reproductive season, agents are continuously iteroparous producing at most one offspring per round. Children have a starting ASI proportional to the parental ASI and the breeding parent pays from its own ASI in order to reproduce. Both sexual and asexual reproduction are available: this is biologically realistic as reference to snails will show. When sexual reproduction is chosen the mate is chosen randomly from mature agents in the immediate population, ie. agents occupying the same lattice space in the environment. Only one agent make the ASI payment to breed but both gain from RRV payoff over subsequent rounds. This too is biologically justified.

The agent has inputs relating to the surrounding environment and aspects of its own state. It has an age related input, this is present to allow life-history details to be introduced, and an overall input reflecting its present state of fitness arrived at simply by summing ASI and RRV. So far as environmental information is concerned and agent can sense the type and density of flora present in its immediate vicinity (the section of grid it currently occupies) and the type of flora present immediately North, South, East and West.

An agent has seven available options for action, these are the primitive behaviours that more complex strategies may be constructed from. There are four movement options - East, West, South, North. Eating and breeding (both sexual or asexual) options exist, eating carrying twice the weight of other options. Finally, it may chose to do nothing. An agent is restricted to one action only per round.

6 Overall System Behaviour

As previously stated the aim of the system is not to use evolution to achieve optimality in an agent, but rather to allow for the emergence of basic levels of adaptation among large populations of agents. At present our investigations are centred on broad aspects of system and agent behaviour.

6.1 Population dynamics and agent–resource interaction

Figs. 6 and 7 show typical population dynamics and population resource relationships. The oscillations in population level are not typical of most ecological systems but reflect the sort of dynamics exhibited by Nicholson's blowflies and the extremes of r-selection where predation has little impact. These oscillations would be damped by predation but the main cause is that inter-agent competition in the system takes the form of *scramble* competition [Nic54].

The cycling of resource and agent levels is exactly what we would expect from an ecosystem operating under similar constraints to the Herby system. The lynx and hare data from the Hudson Bay Company [Sig93] is another example.

I order to cover a fuller set of natural systems we are introducing simple mechanisms of *contest* competition. This will allow us to further refine agent-agent interactions to allow for differential resource acquisition.

6.2 Adaptation

Prior to sampling the population we investigated the general trends in agent behaviour over time. In order to establish a baseline for agent behaviour an agent identical in all features to that considered above has been implemented. However, this control agent has no means of adaptation, decisions are made on a purely *Monte Carlo* basis. The proportions of the agent population exhibiting certain behaviours over time is plotted in fig 8, extended runs have exhibited no trends away from the basic proportions illustrated. Naturally, this is exactly as it should be as there is no adaptive component in this system. Conversely, running a simulation with adaptive agents in the same environment subject

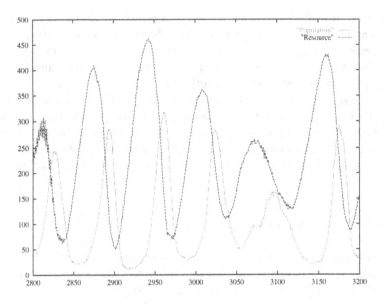

Fig. 8. Lotka-Volterra Type Cycling

to identical initial conditions produces divergence in the relative frequencies of basic behaviours, this is illustrated in fig 9. Thus we found that adaptive agents developed more biologically meaningful (relevant) behaviours than the controls. Interestingly, the same environment with an alternative seed value for the rule base produces the dynamics shown in fig 10, the transition is stepped rather than smooth.

Of particular interest was whether or not strategies would emerge from combinations of the basic behaviours, and if so would they constitute ESSs [Smi76]. At early stages in the simulation the distribution of the basic behaviour each timestep took on overall trends. The proportion of agents eating increased and that of agents moving, an expensive behaviour, decreased. The proportion doing nothing, a cheap but ultimately doomed selection decreased markedly and remained low with occasional resurgences. As the simulations progressed the were long stretches of quite marked periodicity, these often coincided with the with the sort of behavioural distributions illustrated in Fig 11. Here the general trend in the population is towards moving or eating alternately. This is an intuitively good strategy as it allows areas to be exploited then a migration to less exploited areas. However, this strategy is not an ESS for the agents within our system. An ESS must be uninvadable by an alternative strategy, these periods of behavioural synchronisation disappeared after many hundreds of timesteps only to emerge once again later.

It was also interesting to observe the incidence of the component directions of movement. The simulation plotted here utilised a toroidal environment comprising thirty-six areas. Two different resources, one considerably better than the other were distributed in four 3x3 diametrically opposed blocks. As the en-

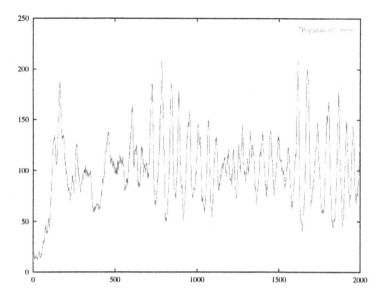

Fig. 9. Boom and bust population dynamics

vironment was relatively small, an agent could traverse it in six timesteps, this arrangement was chosen to ensure that an agent could not find fresh food merely by moving in one direction, some appreciation to the surrounding areas was necessary. Figures 14 to 17 show how the proportions of directional components to change over time. It is important for agents to move along both North-South and East-West axes. However, the size of the environment means that the actual direction with these axes is unimportant so long as they are well represented. I appears that drift may cause the balance between components to change, and even the near elimination of the East 'allele', so long as the axis itself is represented.

Figures 18 and 19 show that doing nothing becomes a very rare behaviour. 20 and 21 show average age and average fitness during periods of low population oscillations (Fig 20) and high oscillations (Fig 121). The variation in average age is consistent with boom and bust dynamics in r-selected populations.

All the data included here came from a simulation with an initial population of ten agents. The agents were instantiated in pairs with randomly generated rulesets, consequently each rule set had two representatives in the initial population. Early experimentation has confirmed the expectation that such 'heterogeneous' starting populations had a higher chance of establishing viable system populations in the long run.

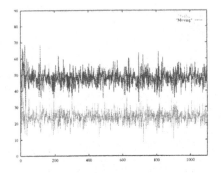

Fig. 10. Behavioural frequencies in a population of random agents

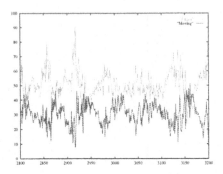

Fig. 13. Synchronisation developing

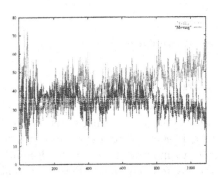

Fig. 11. Behavioural divergence in a population of adaptive agents

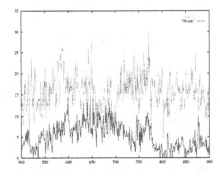

Fig. 14. Incidence of North and South movement, 500–900 timesteps

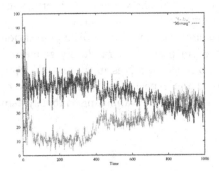

Fig. 12. Behavioural frequencies with an alternative seed value

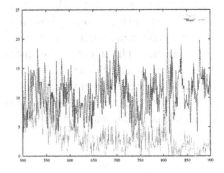

Fig. 15. Incidence of East and West movement, 500–900 timesteps

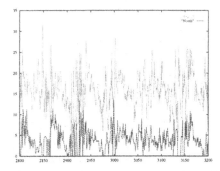

Fig. 16. Incidence of North and South movement, 2800–3200 timesteps

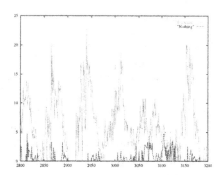

Fig. 19. Relative incidence of feeding and inaction between 2800 and 3200 timesteps

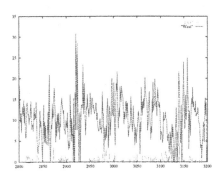

Fig. 17. Incidence of East and West movement, 2800–3200 timesteps

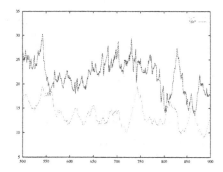

Fig. 20. Average age and average 'fitness' between 500 and 900 timesteps

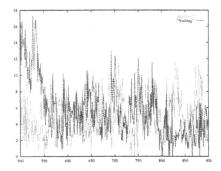

Fig. 18. Relative incidence of feeding and inaction, 500–900 timesteps

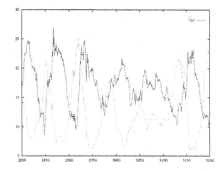

Fig. 21. Average age and average 'fitness' between 2800 and 3200 timesteps

7 Concluding Remarks

This system has been conceived and designed to simulate essential aspects of real ecosystems. Spatial heterogeneity, resource behaviour and interactions have been directly inspired by real ecology. At present the aspect of the system that is less general than had been desired is the *scramble* competition that exists between agents. In the majority of systems animals appear to enter into contests for food where the demand is greater than supply, some individuals take more than others and this variation is an important factor in population dynamics [Lom78]. The variation between agents should be extended to incorporating variation in the amount of resource actually acquired during the process of eating. It is clear that selection occurs across different levels and timescales. Selection for rules and combinations of rules occurs from generation to generation as mutation and crossover produce new rule bases and the effects of differential survival and reproduction affect the representation of good and bad rules and combinations in the population of agents. The rule discovery and credit assignment algorithm effectively selects substrings from with S over somatic time. However, this does have an effect over evolutionary time as those rules that provide good substrings are more likely to be passed on down the generations.

8 Acknowledgements

We would like to thank Prof. Geoff Parker FRS for his interest and support during this project. This work was funded by the Engineering and Physical Science Research Council.

References

[AL91]   D. Ackley and M. Littman. Interactions between learning and evolution. In Farmer, Langton, Rasmussen, and Taylor, editors, *Artificial Life II*, pages 488–509, Reading (Mass.), 1991. Addison-Wesley.

[Bal96]   J. M. Baldwin. A new factor in evolution. *The American Naturalist*, 30:441–451 and 536–533, 1896.

[Beg85]   M. Begon. A general theory of life-history variation. In R.M. Silby and R.H. Smith, editors, *Behavioural Ecology - Ecological Consequences of Adaptive Behaviour*. Blackwell, 1985.

[BGH90]   L. Booker, D.E. Goldberg, and J.H. Holland. Classifier systems and genetic algorithms. *Artificial Intelligence*, 1990.

[CL81]   G. Coughley and J.H. Laughton. Plant-herbivore systems. In R.M. May, editor, *Theoretical Ecology*, pages 132–166. Blackwell, Oxford, 2 edition, 1981.

[CMB95]   F. Cecconi, F. Menczer, and R. K. Belew. Maturation and the evolution of imitative learning in artificial organisms. *Adaptive Behaviour*, 4(1):29–50, 1995.

[CWFL93]   R. Constanza, L. Wainger, C. Folke, and K. G. M. Ler. Modeling complex ecological economic systems: Toward an evolutionary, dynamic understanding of humans and nature. *BioScience*, 43:545–555, 1993.

[DC89]   L. Darden and J. A. Cain. Selection type theories. *Philosophy of Science*, 56:106–129, 1989.

[DKP96]   P. Devine, G. Kendall, and R. Paton. When herby met elvis – experiments with genetics based learning systems. In V. J. Rayward-Smith, editor, *Modern Heuristic Search Methods*, pages 275–292. J. Wiley and Sons, 1996.

[DP97]   P. Devine and R. C. Paton. Herby, an artificial evolutionary ecology. In *Proceeding of ICEC97*. IEEE Press, 1997.

[FJ94]   S. Forrest and T. Jones. Modeling complex adaptive systems with echo. In R. J. Stonier and X. H. Yu, editors, *Complex Systems: Mechanisms of Adaptation*, pages 3–21. IOS Press, 1994.

[FM94]   R. French and A. Messinger. Genes, phenes and the baldwin effect. In *Artificial Life IV*. MIT Press, 1994.

[HDP84]   M. Huston, D. DeAngelis, and W. Post. New computer models unify ecological theory. *Bioscience*, 38(10):682–691, 1984.

[Hei94]   D. Heibeler. The swarm simulation system and individual-based modeling. In *Decision Support 2001: Advanced Technology for Natural Resource Management*, 1994.

[Hol92]   J. H. Holland. *Adaptation in Natural and Artificial Systems*. MIT Press, 1992.

[Jud94]   O. P. Judson. The rise of the individual-based model in ecology. *Trends in Ecology and Evolution*, 9:9–14, 1994.

[Lev66]   R. Levins. The strategy of model building in population biology. *American Scientist*, 54:421–431, 1966.

[Lom78]   A. Lomnicki. Individual differences between animals and the natural regulation of their numbers. *Journal of Animal Ecology*, 47:461–475, 1978.

[MG94]   J. A. Meyer and A. Guillot. From sab90 to sab94: Four years of animat research. In Cliff, Husbands, Meyer, and Wilson, editors, *From animals to animats 3. Proceedings of the third international conference on simulation of adaptive behaviour*. MIT Press, 1994.

[Mon67]   J. Monro. The exploitation and conservation of resources by populations of insects. *Journal of Animal Ecology*, 36:531–47, 1967.

[Nic54]   A. J. Nicholson. An outline of the dynamics of animal populations. *Australian Journal of Zoology*, 2:551–598, 1954.

[Pia74]   E. R. Pianka. *Evolutionary Ecology*. Harper and Row, New York, 1974.

[PNC92]   D. Parisi, S. Nolfi, and F. Cecconi. Learning, behaviour and evolution. In F. J. Varela and P. Bourgine, editors, *Towards a practice of autonomous systems*, pages 207–216, Cambridge (Mass.), 1992. MIT Press.

[Pri84]   P. W. Price. Alternative paradigms in community ecology. In P. W. Price, C. N. Slobodchikoff, and W. S. Gaud, editors, *A New Ecology, Novel Approaches to Interactive Systems*, pages 353–383. J. Wiley and Sons, 1984.

[Sig93]   K. Sigmund. *Games of Life*. Oxford University Press, 1993.

[Smi76]   J. Maynard Smith. Evolution and the theory of games. *American Scientist*, 64, 1976.

[Wie84]   J. A. Wiens. Resource systems, populations and communities. In P. W. Price, C. N. Slobodchikoff, and W. S. Gaud, editors, *A New Ecology, Novel Approaches to Interactive Systems*, pages 397–346. J. Wiley and Sons, 1984.

[Wil90]   S. W. Wilson. The Animat path to AI. In *From Animals to Animats - Proc 1st Int Conf on Simulation of Adaptive Behaviour*, 1990.

Modelling Bounded Rationality
Using Evolutionary Techniques

Bruce Edmonds and Scott Moss

Centre for Policy Modelling, Manchester Metropolitan University,
Aytoun Building, Aytoun Street, Manchester, M1 3GHG, UK.
http://www.cpm.mmu.ac.uk

Abstract. A technique for the credible modelling of economic agents with bounded rationality based on the evolutionary techniques is described. The genetic programming paradigm is most suited due to its meaningful and flexible genome. The fact we are aiming to model agents with real characteristics implies a different approach from those evolutionary algorithms designed to efficiently solve specific problems. Some of these are that we use very small populations, it is based on different operators and uses a breeding selection mechanism. It is precisely some of the "pathological" features of this algorithm that capture the target behaviour. Some possibilities for integration of deductive logic-based approaches and the GP paradigm are suggested. An example application of an agent seeking to maximise its utility by modelling its own utility function is briefly described.

1   Introduction

The purpose of this paper is to report on an approach to simulating economic agents such that their behaviour matches their known characteristics. The approach taken is to introduce characteristics of bounded rationality and learning into these simulations using evolutionary techniques.

By using an approach to modelling learning that is close to that used in genetic programming (GP) [10, 11], we open up a new range of possibilities in the credible modelling of such boundedly rational agents, where an agent has a population of candidate beliefs (or models) of its environment which evolve as it learns. This contrasts in several respects from agent modelling approaches that use "crisp" logic-like beliefs, or approaches that only involve some inductive learning. In particular multiple and frequently inconsistent beliefs are held as a resource for future model development. However, despite this contrast this approach supports integration of such a style of learning with deductive mechanisms.

2   Modelling Boundedly Rational Economic Agents

If you seek to model real economic agents then, unless you make some very sweeping assumptions, the entities in your software model will also need the broad characteristics of the real agents. This is in contrast to traditional economics where, by and large, the *agency* nature of the agents is ignored, in favour of trying to capture their behaviour *en masse*.

Thus the purpose of an agent in such a model is different from either agents that are designed with a particular purpose in mind (e.g. [5]) or for exploration of the most effective and flexible algorithm for a set of problems. In such modelling one seeks for as much veracity as is possible given the usual limitations of time, cost and technique and one does not necessarily look to design them to be efficient, general, or consistent in their beliefs.

In particular we are interested in agents who:
- do not have perfect information about their environment, in general it will only acquire information through interaction with its environment which will be dynamically changing;
- do not have a perfect model of their environment;
- have limited computational power, so they can't work out all the logical consequences of their knowledge [18];
- other resources, like memory are limited (so they can't hold large populations of models);

In addition to these bounds on their rationality we also add some other observed characteristics of real economic agents, namely:
- the mechanisms of learning dominate the mechanisms of deduction in deciding their action;
- they tend to learn in an incremental, path-dependent [1] (or "exploitative") way rather than attempting a global search for the best possible model [16];
- even though they can't perform inconsistent actions, they often entertain mutually inconsistent models and beliefs.

The fundamental difference between these agents and, say, logic-based agents, is that the updating of internal belief structures is done in a competitive evolutionary manner using a continuously variable fitness measure rather than in a "crisp" consistency preserving manner. This is appropriate in situations of great uncertainty caused by a rationality that is not able to completely "cope" with its environment but is more restricted in its ability.

3   The Agent Architecture

For the above reasons we have developed a paradigm of modelling the learning that such agents engage in, as itself a process of modelling by the agents. For more on this framework see [13].

Although economic agents primarily develop though a process of incremental learning they also use some deductive procedures. In real economic agents these processes may be arbitrarily mixed as well as developed and abstracted over different layers of an organisation. Here we will only look at a model which effectively separates out learning and deduction and comes from an essentially unitary agent structure.

The agent works within a given *a priori* body of knowledge (e.g. accounting rules). The agent may well make deductions from this in a traditional way and apply these to the current hypotheses. This body of *a priori* knowledge may also determine the syntax of the models the agent starts with, its principal goals, default actions, fitness functions and the operations to be applied to its models. Typically much of this a priori knowledge can be made implicit in the syntax of the genome (which is the approach we have tended to take).

The agent here has many models of its environment. Once started the agent incrementally develops and propagates these models according to a fitness function which is based on its memory of past data and effects of its actions as well as the complexity and specificity of its models. It then selects the best such model according to that measure. From the best such model and its goals it attempts to determine its action using a search-based, deductive or quasi-deductive mechanism. It then takes that action and notes the effects in the environment for future use. The setup is illustrated below in figure 1.

The development of these models (i.e. the learning) is modelled by an evolutionary process on this population of internal models (similar to that described in [4]). Important restrictions on such agents include the fact that it may have only limited information gained as the result of inter-action with its environment and that any action costs it so that it can not indulge in an extensive exploratory search without this being weighed against the benefit being gained (this is especially true given the course temporal graining of typical economic simulations).

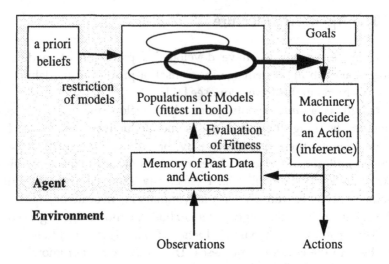

Figure 1: Basic Structure of a Simplified Economic Agent

4   The GP Paradigm

An important special case of the above approach to learning is where the range of operations includes selection and some mechanism for variation, i.e. an evolutionary algorithm. In particular the paradigm of GP is particularly appropriate, due to the structure of the genome. These techniques, however, can not be blindly applied. For example, the efficiency of the learning process is only a secondary concern when seeking to model economic agents by their software cousins, but many of the other features of this approach for modelling learning in an economic agent *are* appropriate, namely:

- the population of programs can represent a collection of multiple, competing models of the world with which it is concerned;
- there is always at least one maximally fit individual model that can be used to react to events and from which appropriate deductions can be made;
- the models are incrementally developed by the learning mechanism;
- the fitness measure can be tailored to include aspects such as cost and complexity as well as the extent of the agreement with known data;
- the language of representation of the models can be fairly general and expressive, e.g. logical expressions.

5   Adapting the GP Paradigm

There are several possible ways of using evolving populations to simulate a community of economic agents:

1. each member of the evolving population could represent one agent;
2. each agent could be modelled by a whole evolving population;
3. the whole population could be modelled by the whole evolving population but without an individually intended agent <-> gene correspondence.

Method (1) has been used in several models of agents which evolve (e.g. [7, 19]), here the genetic development has nothing to do with the nature of an agent's cognitive processes but helps determine its goals or strategies. Method (3) above is the most popular in economics (e.g. [1, 3]), but unless such a model predicts pertinent properties of real populations of agents it represents a bit of a fudge, and means that the observable behaviour and content of individual entities in the model do not have a clear referent in what is being modelled. This makes it far less useful if one wants to use such models to gain a detailed insight into the internal dynamics of populations. Method (2) actually addresses the cognitive process as the agent corresponds to a population of mental models. This has been done before in a limited way in [15], but here agents have a fixed menu of possible models which do not develop.

In using the evolutionary paradigm in this sort of modelling we tend to:

- represent the agent by a whole evolving population - each gene corresponding to one of its alternative models (this is the approach taken in the example in Section 7);
- populations of agents are thus modelled as populations of evolving populations (i.e. populations of populations), with an intended agent to evolving population correspondence (e.g. [14]);
- give the agents only small populations of models, representing limited memory;
- base the fitness function on a combination of its error compared to past data, size of model and its predictivity (precision and range of applicability);
- restrict the variation operators to such as generalisation, specialisation, averaging, combining and mutating;
- and give them only a limited inferential ability to use its best model to choose its action.

This paradigm needs to be integrated with an agent-based approach and adapted to relate to credible models of economic agents. In particular the cross-over operator is somewhat arbitrary when simulating the development of models in economic agents (although undeniably efficient). This also introduces a globality to the search which is unrealistic.

In the example application presented below we use a process of combining old models together as branches from a new node and introducing randomly generated small new models. This produces more realistic results, for example it allows for better fitting by parameterisation.

6 Possible integration with Logic-based Agent Architectures

The structure of the agent described above and the GP style of chromosome allow for some integrations of the evolutionary learning mechanism and agents based on logic based inference mechanisms (e.g. the BDI framework of). This can occur when the chromosomes representing the internal mental models of the agent represent logical expressions. Given this there are two main possibilities:

1. The population of genes can be constrained to those that are logically consistent with a set of a priori knowledge expressed as logical expressions within some formal logical framework;
2. The inference of possible actions could be done by using a logical framework to infer the best action from its best model and its goals;

It would not be appropriate to constrain the population of internal models so that they were consistent with each other or consistent with its goals as these represent the competing partial beliefs of the agent about its world.

7 An Application - a model of utility learning

A simple application of the above approach is that of an economic agent that seeks to maximise its utility by dividing its spending of a fixed budget between two goods in each time period (what it does not spend on one good it spends on the other).

Unlike classical economic agents, this one does not know its utility function (even its form) but tries to induce it from past experience. It only gets information about the utility of a particular spending pattern by actually trying it. The agent wants to get the most utility from its spending. It will not speculate with alternative spending patterns merely to learn more about the utility curve.

To do this it attempts to model its utility with a function represented by a GP type chromosome using +, -, *, /, max, min, log, exp, average, "cutbetween" (a three-argument function which takes the second value if the first value is less than 1 and the third value thereafter) as branching nodes, and a selection of random constants and variables representing the amounts bought of the two products for the leaves. Thus the chromosome

[average
　　[divide
　　　　[amountBoughtOf 'product-2']
　　　　[constant 2.3]]
　　[constant 0.5]]

would predict that the utility gained would be

$$\frac{(x/2.3) + 0.5}{2}.$$

Where x is the amount spent of product 2.

The fitness function is based on the RMS error of the prediction of a model compared to the actual utility gained over past spending actions. This is modified by a slight parsimony pressure in favour of shallower chromosomes and a bias in favour of chromosomes which mentioned more distinct variables based on the among bought (a rough measure of specificity - called "volume" in [13]). Only the fittest half of the population is retained each generation, so that this is a kind of selective breeding algorithm and does not use fitness proportionate random selection (thus is has some similarities to evolutionary programming [6]).

Each time period the agent:

1. carries over its previous functional models;
2. produces some new ones by either combining the previous models with a new operator or by growing a small new random one;
3. evaluates its current models using past data;
4. selects the best models in terms of fitness for survival;
5. it finds the fittest such model;
6. it then preforms a limited binary search on this model to find a reasonable spending pattern in terms of increasing its utility;
7. finally it takes that action and observers its resulting utility.

This model was realised in a language called SDML (Strictly Declarative Modelling Language) - a language that has been specifically developed in-house for this type of modelling. This is a declarative object-oriented language with features that are optimized for the modelling of such *economic, business and organisational* agents [9, 20].

Limiting the depth of the models created to 10, We preformed 10 runs over 100 time periods for each type of agent. The three types were characterised by the memory they were given and the number of new models they created each

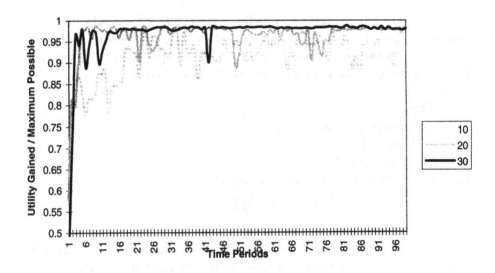

Figure 3: Utility Ratio Achieved for Agents with Different Memories, Averaged over 10 Runs

To give a flavour of the sort of models these agents develop, in run 1 of the 30-memory agent batch the agent achieved the following model by date 75:

```
[average
    [[divide
        [[add [[constant 1.117] [amountBoughtOf 'product-2']]]
        [average [[amountBoughtOf 'product-2'] [constant 4.773]]]]]
    [min
        [[amountBoughtOf 'product-2']
        [cutBetween
            [[average [[amountBoughtOf 'product-2'] [constant 4.773]]]
            [constant 1.044]
            [add [[constant 1.117] [amountBoughtOf 'product-2']]]]]]]]].
```

The extent of the fit learnt by the agent is shown in figure 4.

time period: respectively 10, 20 and 30. We call these 10-memory, 20-memory and 30-memory agents, they represent agents with different bounds on their rationality. The results were then averaged over these 10 runs.

The first graph shows the (RMS) error of the agent's best model of the utility function compared with the actual function (figure 2). It shows a great improvement between the 10-memory agent's and 20-memory agents, but only a marginal improvement between 20 and 30-memory agent's, suggesting the existence of a sort of minimum capacity for this task.

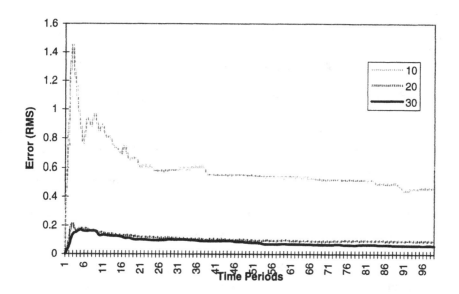

Figure 2: Error in Agent's Best Model for Different Memories, Averaged Over 10 Runs

When you look at the utilities achieved by the agents with different memories (figure 3), you see that a memory capacity (above 10) does not significantly increase the average utility over time, but it *does* dramatically effect the reliability of the utility it gains. If this were a firm with the utility being its profits, this reliability would almost as important as its average profit level.

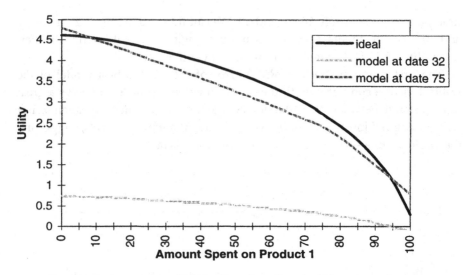

Figure 4: Learnt vs. Actual Utility Functions, Run 1 of 30-memory Agents

The purpose of this simulation is *not* to be an efficient maximiser of utility, but to model economic agents in a more credible way. It will only be vindicated (or otherwise) when compared to real economic data. However, the model does show traits found in the real world. For example, one phenomenon that is observed is that agents sometimes get "locked" into inferior models for a considerable length of time (as in [2]) - the model implies an inferior course of action, but this course of action is such that the agent never receives disconformation of its model. Thus this remains its best model in terms of the limited data it has, so it repeats that action. If, for example, some consumers find a satisfactory brand at an early stage in the development of their tastes and then they may never try any others - their (limited) experience will never disconfirm their model of what would give them most satisfaction, even when they would like other brands better.

Other related applications have included a model of intelligent price fixing in Cournot Duopoly tournaments [12], and a model of emerging markets where the agents are simultaneously building models of the economy they inhabit (and mutually create) [14].

8   Discussion

Such modelling using evolutionary techniques, where there is an explicit one-one correspondence between items modelled and the genes in the population typically deal with very small populations (in evolutionary terms).

In the example above we had populations of mental models as small as 10. Most of the models of abstract evolutionary algorithms deal only with large populations (many assume an infinite population for formal purposes). The behaviour of small populations may be pathological from the point of view of an efficient search mechanism, but here we have different goals in using evolutionary algorithms. It is precisely the pathological aspects of the process that capture the qualitative behaviour observed: sharp path-dependence, lock-in, exploitative search, a large spread of behaviours between different populations and limited overall optimization.

Also it is not always the case that the usual genetic operators are very efficient in such small populations. It is known that selective breeding can work well with small populations [8]. In addition (in the example above) we found that a traditional GP mixture of tree-crossover and propagation did substantially worse than that of combining together old models, generating small new random ones and propagation. This is important as the mechanism chosen has to be credible for realistically small populations of mental models. Much work needs to be done to understand the evolutionary dynamics of small populations.

9   Future Work

Research into this style of modelling agents is at a very early stage. Future work is almost unbounded. There some more immediate shortcomings of this approach that we intent to focus on soon. In no particular order these include:
- the introduction of intentions and planning;
- the development of techniques to evolving logical expressions, representing beliefs;
- the integration of the evolutionary module, with existing logic based approaches as described above;
- the investigation of the effects of different syntaxes (as exhorted by [17]);
- the increasing formalisation of the structure;
- the investigation of the effects of different genetic operators in very small populations

References

1. Arifovic, J. (1994). Genetic Algorithm Learning and the Cobweb Model, Journal of Economic Dynamics and Control 18, 3-28.
2. Arthur, W. B. (1995). Increasing Returns and Path Dependence in the Economy. University of Michigan Press, Ann Arbor, MI.
3. Axelrod, R. (1984). The Evolution of Cooperation, Basic Books, New York.

4. Dennett, D.C. (1995). Consciousness Explained. Philosophy and Phenomenological Research, 53, 889-892.
5. Bonasso, R. P.; Kortenkamp, D.; Miller, D. P.; Slack, M. (1995). Experience with an Architecture for Intelligent Reactive Agents. ATAL'95 at IJCAI'95, Montreal. Published in (1996), Lecture Notes in Artificial Intelligence, 1037, 187-202.
6. Fogel, D.B. (1994) Evolutionary Programming an introduction and some current directions. Statistics and Computing, 4, 113-129.
7. Holland, J. H. (1992). Adaptation in Natural and Artificial Systems, 2nd Ed., MIT Press, Cambridge, MA.
8. Mühlenbein, H.; Schlierkamp-Voosen, D. (1993). The science of breeding and its application to the breeder genetic algorithm (BGA). Evolutionary Computation, 1, 335-360.
9. Edmonds, B.; Moss, S.; Wallis, S. (1996). Logic, Reasoning and A Programming Language for Simulating Economic and Business Processes with Artificially Intelligent Agents, AIEM96, Tel Aviv, 1996.
10. Koza, J. R. (1992). Genetic Programming: On the Programming of Computers by Means of Natural Selection. MIT Press, Cambridge, MA.
11. Koza, J. R. (1994). Genetic Programming II: Automatic Discovery of Reusable Programs. MIT Press, Cambridge, MA.
12. Moss, S.; Dixon, H.; Wallis, S. (1995). Evaluating Competitive Strategies. Intelligent Systems in Accounting, Finance and Management. 4, 245-258. Also available at URL: http://www.cpm.mmu.ac.uk/cpmrep02.html
13. Moss, S.; Edmonds, B.; (1994) Modelling Learning as Modelling, CPM Report 3, Centre for Policy Modelling, Manchester Metropolitan University. Available at URL: http://www.cpm.mmu.ac.uk/cpmrep03.html
14. Moss, S.; Kuznetsova, O. (1995). Modelling the Process of Market Emergence. MODEST (Modelling of Economies and Societies in Transition), Warsaw, 1995.
15. Palmer, R.G. et. al (1994). Artificial Economic Life - A simple model of a stockmarket. Physica D, 75, 264-274.
16. Penrose, E.T. (1959) The theory of the Growth of the Firm, Blackwell, Oxford.
17. Russell, S.J.; Grosof, B.N. (1990): A Sketch of Autonomous Learning using Declarative Bias. In: Machine Learning, Meta-Reasoning and Logics. (Eds: Brazdil, P.B.; Konolige, K.) Kluwer Academic, Boston, 19-53.
18. Simon, H.A. 1972. Theories of Bounded Rationality. In McGuire, C.B.and Radner, R. (eds.) Decision and Organization. North-Holland.: Amsterdam.
19. Vriend, N.J. (1995). Self-organization of markets: an example of a computational approach. Computational Economics, 8, 205-232.
20. Wallis, S.; Edmonds, B.; Moss, S. (1995). The Implementation and Logic of a Strictly Declarative Modelling Language (SDML). ES'95, Cambridge, UK.
21. Wooldridge, M. (1995). Time, Knowledge, and Choice. ATAL'95 at IJCAI'95, Montreal. Published in (1996), Lecture Notes in Artificial Intelligence, 1037, 79-96. Available at URL: http://www.doc.mmu.ac.uk/STAFF/mike/atal95.ps

The Abstract Theory of Evolution of the Living

V.L. Kalmykov

Institute of Cell Biophysics, Russian Academy of Sciences,
Pushchino, Moscow Region, 142292, Russia
E-mail: kalmykov@ibfk.nifhi.ac.ru

Abstract. The paper is the result of attempts of the biologist to formulate the logic of evolution of the living for its probable use in Artificial Intelligence, especially for purposes of Evolutionary Computing and Artificial Life. The logic has been fulfilled as the abstract theoretical model, which integrally reflects the universal functional algorithm of evolution of the living. The essence of the method was the invention of an ideal objects (a creative synthetic definitions) that are still absent for a sufficient understanding of evolution. The generalized definitions of the following notions are given:

Endergonic structures; Types of the endergonic structures of order;
Information; Complete entropy; Level of organization;
Functional intraclosures (organisms, organizations);
Criterion of the life evolution direction (Parameter of the comparative evolutionary progress of the structures);
Creation; Life; The invariant cognitive cycle; ...

Introduction.

This work presents a generalized theoretical description (functional scheme) of the living. The task is close to the questions of what life is and why it is organized so. Being traditional for the theoretical biology, these questions are at present especially urgent in connection with the problems of development of computers, robots and cyberspace. In the latter cases the development of computers with soft- or/and hardware, based in essence on the logic of organization and evolution of the living, is meant. Besides, for a valuable and justified presentation of knowledge about the living in computers its generalized (universal) functional description is necessary. This work presents such a generalized theoretical description of the living. It includes elements of axiomatic approach and is physically interpretable. A mathematical groupoid of elementary operations, which are not reduced to each other, is suggested as a fundamental notion for the living world. The mathematical groupoid is the functional invariant of the living organization. The regulations for mechanisms of integral closure of elementary operations onto each other in the course of biological self- organization are examined as well.

The suggested functional invariant of organization and evolution appears to be fundamental not only for biological objects themselves but also for any organizational levels of the living.

Why do I try to use a group theoretical approach? As I understand, it is the only one way to formulate the main point of organization, behaviour and evolution of the living for science and technological use. The low of organization of integral structure and the structure symmetry is one and the same. **The structure symmetry** *is the most high automorphism group of the structure* [1].

The earlier variants of the results obtained in this direction were prepublished [2-4].

The main statements, notions and interpretations.

1. **The set of compatible structures (M)** is the basis of the living. *These structures are formed by fixation of free environmental energy in their structure and, as a result, they are able to do some work.* Such structures will further be called **endergonic** ones. **Compatibility of these structures** *is structural homomorphism in character, i.e. they have a fundamental unity of specific morphological arrangement. In consequence, there is some easiness of their interaction up to the possibility of a reciprocal transformation.*

2. The environment, as an initial source and a final drainage receiver of substance and/or energy, is necessary for existence of the living. The environment is assumed to afford some interval of conditions for realization of optimal kinetic stability of structures of the living.

3. Endergonic structures of the set M possess such a vast structural variety (polymorphism), that they are capable of establishing **ten pairs of simple mutually opposite functional relations between each other, i.e. between their constituents and the environment (set R)** (SEE Table 1). This set of functions is basically invariant for the living. The notion "function" used here is analogous to its use in the work by G.A.Chauvet [5]. Stressing the orientation and asymmetry of the notion "function", the word "operation" is used in this work as a synonym.

Table 1. Set R includes ten pairs of mutually opposite elementary operations (with substance, energy and information) on the set of compatible endergonic structures of set M that underlie the living

Direct operations	Reverse operations
1. Identification	1'. Identification
2. Right-hand mirror reflection	2'. Left-hand mirror reflection
3. Change of position in space	3'. Revertion of position in space
4. Transformation of configuration	4'. Restoration of configuration
5. Increase of number of elements	5'. Decrease of number of elements
6. Provision of inlet	6'. Provision of outlet
7. Connection	7'. Isolation
8. Switching-on, i.e. mediating initiation of a specific action of the operation object, affected by a definite way	8'. Switching-off
9. Inflowing	9'. Outflowing
10. Concentrating	10'. Scattering

The operations of set R and their combinations entirely cover all kinds of relations that are obligatory for emergence and a stable existence of endergonic structures of the set M. The operations are equally performed with substance and energy and information.

Operations R generate the **mathematical groupoid G** over all possible combinations. *The proofs are:*

1. In case of combinations (unlike permutation) the sequence of operations is not significant, hence the performance of properties of associativity appears.

2. All combinations of operations belong to one groupoid. This follows from the conditions of specifying this set, in combinations of which all possible changes in the structure are embedded.

3. There is the only common unit, which is the operation of identification.

4. There is a reverse element for each element (SEE Table 1).

Conceivably the groupoid G might represent a strict group, but there is necessary to look more closely at this suggestion.

Self-organization *is a spontaneous emergence of the structures of order in the course of spontaneous processes.* **The endergonic structures of order** *are in principle thermodynamically instable kinetic stabilities.* During self-organization the spontaneous transitions from one structure of order to another are conditioned by thermodynamic instability (nonequilibrium).

The possible **types of order of the endergonic structures** are as follows:

1. *The static ones.* For instance, organic molecules (including macromolecules) and their crystals.

2. *The informationally unmediated stationary structures.* They are dynamic structures existing due to an informationally unmediated return to the initial position (state, form). For example, dissipative autocatalitic structures of the Beloussov - Zhabotinsky reaction type [6], whirlwinds, rivers (permanent stations based on the water circulation)...

3. *The informationally mediated stationary structures.* They are dynamic structures existing due to an informationally mediated return to the initial position (state, form). The kinds of such automorphic processes are: reproduction, adaptive behaviour, recovery (regeneration, repair).

As it is seen from the points listed, any endergonic structure of order can be characterized by its specific group of symmetry. In particular, this appears from the fact that the set of transformations, which make the structure return to its initial position (state, form), is just one of the definitions of a group of symmetry. Groupoid G characterized here is common to all the possible endergonic structures.

Let us consider two neighbouring levels of the structure organization: the structure itself and its substructures of the first below-lying level. *When examining the structure (a complex of interacting substructures) as a single whole (as if "from outside"), we are speaking about the* **macroapproach.** Here the inner substructures (microlevel) are ignored, and generalized characteristics of the state only are relevant. The generalized characteristics, like free energy, symmetry and entropy, allow us to speak about the structure transformations (transitional structures of order).

In case of the **microapproach** *the structure is supposed to be examined from inside, and behavioral characteristics of the substructures(microstructures) are relevant.*

In the **macroapproach** *the notion "space of possible (virtual) transitional states of structures of order" is used.*

In the microapproach we use the notion "space of possible behavioral forms of the substructure (microstructure) of the first below-lying level of structure organization".

There are certain **criteria of the direction of spontaneous autonomous transformations of macrostructures,** i.e. a spontaneous behavior of microstructures within an isolated macrostructure is only "permitted" when:

- (1) free energy of the macrostructure (its capability of doing some work) decreases;
- (2) entropy of the macrostructure increases;
- (3) symmetry of the macrostructure increases.

Criteria 1 and 2 formulate the 2nd law of thermodynamics. Criterion 3 is a corollary of the general principle of symmetry of physical phenomena formulated by Pierre Curie [7] in 1894. This principle is also developed in the works by A.V Shubnikov and V.A. Kopcik [1,8].

Free energy of the structure is defined as *the ability to do some work autonomously.*

Symmetry of the structure *is characterized by an at most high group of transformations, which leave the structure unchanged.* "Most high" mean "including all possible transformations"

As for entropy, the matter is more difficult. In this work **complete entropy of the isolated structure of order** *is defined as a volume of the space of possible transitional structures of order on their way to equilibrium.* This definition is close to that of absolute entropy by Max Plank [9,10], which does not require any probabilistic ideas and is applicable to physical structures with any number of freedom degrees. The author believe that the conception about space of possible transitional structures of order could be defined as a space of possible transitional groups of the structure symmetry. Hence it appears that complete entropy of the structure is an extensive characteristic of its potential (evolutionary achievable) symmetry, which seems to be rather paradoxical.

To think that the structure is entirely isolated is just idealization. When extreme principles are applied to a real structure, recourse to its mental isolation must be had. In fact, the structure remains open. The mental isolation of the structure is emphasized by means of the notion "independence" of its transformations (behavior). It is most correct to use the formulated extreme principles locally, as a criterion for choosing the direction of spontaneous processes at each concrete point of evolution, but not "on the whole".

Origin and evolution of the living *is self-organization of structures of order.* The complexes of endergonic structures of set M act in the capacity of these structures and carry the combinations of operations of set R. Spontaneous formation of these complexes occur in such a way that the combinations, satisfying the formulated extreme principles, are realized. New self-organized structures of order are formed on the basis of the kinetic stabilities achieved earlier. So, **each step within self- organization** *means an increase of specific contribution of new-emerging structures to realization of the formulated extreme principles, in particular the specific power and/or specific symmetry of the structure continuously increase.* Therefore

the integral criterion of the evolutionary direction *is arising of the specific product of the power of the structure on its symmetry (the mass of the structure is the divisor of the product of its power on its symmetry).*

Direction of evolution is defined as correspondence with this criterion. This criterion is one and the same for direction of evolution and for **direction of progress.**

Let us consider **the stages of emergence of living organisms** on the basis of the set of endergonic structures M. The following **steps of self- organization of endergonic structures leading to origin and evolution of the living** can be discerned:

1) emergence of elementary cycles of catalysis;
2) emergence of cycles of autocatalysis consisting of elementary cycles;
3) emergence of hypercycles [11] consisting of cycles of autocatalysers;
4) formation of functional intraclosures (organisms) over the operations of set R of hypercyclic, autocatalytic, catalytic and simple noncatalytic structures. Two operations of the functional intraclosures finally remain open to the environment: inflowing from sources and outflowing of drains.

1') Beginning of the next iteration of the self-organization. Functional intraclosures (organisms), formed by this process fill up the set of self-organized endergonic structures M. They belong to a higher level of organization. The new level of organization is initial for recurrence of the described self-organized logic etc. The last peculiarity - *the functional isomorphism of organization of the living on any levels of organization is the property of* **the fractal organization of life.**

It may be hypothesized that the algorithm of the cycle of self-organization from 1 to 1', etc. is the invariant not only for organisms and organizations, but and for ideas. In the last case it may be named **the invariant cognitive cycle.**

The level of organization *is a stage in evolution of structures, on which the functional intraclosure of their substructures goes with a principle unity of their specific morphological conformation (with their structural homomorphism).*

The functional intraclosure of organism substructures *makes sense that when interacting with each other, with the environment and other organisms, they provide a kinetically stable existence and development of the organism and, thereby, the existence and development of each other.* It should be noted that such a functional intraclosure is realized in the limit of all organisms and factors of the environment and inner medium, i.e. within the life process as a whole (biosphere). A living organism is therefore both a functional intraclosure and a partial functional closer of the life process factors to a single whole. In the last sense the *biosphere* (the life process as a whole) is interclosure of all of the intraclosures (organisms) in the united intraclosure. An automatic forced selection of alternative combinations of the elements under consideration goes in the direction satisfying the formulated extreme principles. Information mediating the selection of these behavioral forms appears at the points where alternative behavioral forms (combinations of dissipation flows) are equally probable or realization of hardly probable behavioral forms is necessary from the standpoint of satisfaction of extreme principles.

Realization of the selection of the given behavioral form by the structure occurs with the help of controlling substructures making selective steps in potential kinetic barriers, which keep back the dissipation of free energy of the structure and/or selectively lower such potential barriers in accordance with the available information.

The mediating function of information becomes a participator of principle in mutual coordination of self-organized endergonic structures beginning with the stage of emergence of hypercycles.

Information is the central factor determining the stability and the functional efficiency of informationally mediated stationary structures, the living organisms belong to. Hence, the main link in the evolutionary process of the living is functional perfection for obtaining, accumulating, processing and using information.

From our standpoint the physical essence of information and the physical essence of the living are in close interrelation. The well- known theory of optimal coding by Claude Shannon, based on the statistical determination of entropy, is very often called the theory of information. Yet both statistical manipulations by quantity of bits in a file and statistical manipulations by quantity of individuals in a population did not allow simply and distinctly to understand the physical essence of these phenomena un now.

A long search for physical specificity of the living did not permit to relate it either to growth, or to reproduction, or to structural regeneration, or to substance exchange ... These phenomena have been found in crystals and other purely physicochemical structures. Information is the only attribute, which specifically inheres in the living. It is a configuring mediating coordination of all processes, taking place in organism, with each other.

Coordination, organization of all living processes presume the presence of purposefulness that is intrinsic for the living. To our mind, the physical side of this purposefulness consists in that the living is not the direct way of realization of extreme physical principles, as it occurs in the common physics, but the organizationally mediated one. *When considering the living a "black box", then the change in generalized physical characteristics at entrance and exit of this black box will correspond to extreme physical principles.* At the same time, some processes, hardly probable from the standpoint of the common physics, may determinedly occur inside this black box. Determined realization of such hardly probable processes is the consequence of that the living is organized, which is mediated by information. Thus, information fulfills the function of mediating the co-organization of living processes. Life is, in its turn, the way of realizing extreme principles of physics through integral co-organization of behaviour of living elements.

Information *is reflection of a definite trajectory of behavior of the structure in the space of its possible behavioral forms that allows the structure an identical reproduction of the selection made by the structure earlier in its behavior.*

It can be also said that **information** *is an interrelation of events fixed in any way. In the last sense*

information and **reflection of the function** is one and the same.

Information is the central link in the mechanisms of coordination of operations of set R within substructures of organisms, communities of organisms and the whole biosphere. Functionally, information manifests itself in three different forms, being part of three integrating functions: control, reproduction and creation. The three functions integrate elementary operations of set R to a single whole within functional intraclosures.

Control *is a directed change in the probability of realising alternative trajectories of the controlled object behavior.*

Reproduction *is a cycle of the structure transformations under control, which results in emergence of its copy.*

Creation *is a combinatorial process aimed at forming a new type of information mediating the structure behavior control and/or reproduction in higher (in accordance to the extremal principles) evolutionary level. The obtained information, in particular, realizes the process of polymorphic reproduction of the structure and of the substructures (i.e. reproduction of a new type of permutation of elements by introducing and/or removing its elements and/or by changing configurations).*

Reproduction *is the hypercycle of the control cycles and creation is the hypercycle of reproduction cycles. Information, arisen in creation cycle is the base for reproduction of controlling structures.*

Life *is a spontaneous process of combinatorial generation of the groupoid G of functional intraclosures (organisms, organizations) by combining the operations of set R above the set of endergonic structures M.*

I thank Mr. Alexander S. Kharitonov for stimulating, helpful discussions and technical assistance in writing and Mrs. Larisa F. Kun'ieva for help in the electronic submission of the paper.

References

1. Shubnikov, A.V.and Koptsik, V.A.: Symmetry in Science and Art, N.Y. chap. 12 (1974)
2. Kalmykov, V.L.: "The significance of the Theoretical Biology for Biotechnology" (preprint in Russian) Pushchino (1988) 11 pages
3. Kalmykov V.L.: "The Functional Scheme of Organization and Evolution of the Living. The meccano of a biocomputer", SMBnet archives (1995) in directory smb/pubs as two files: <README_VL_Kalmykov_FSOEL> and <VL_Kalmykov_FSOEL.tar.Z.>; files may be got via:
   ftp://ftp.ncifcrf.gov/smb/pubs/README_VL_Kalmykov_FSOEL
   or http://www.iam.ubc.ca/spider/spiros/smb/index.html
4. Kalmykov V.L.: "The Integral Algorithm of Organization and Evolution of the Living Up to Culture - the Possible Instrument for Genetic Programming". In Proceedings of the First Online Workshop on Soft Computing (1996) pp. 284-289 Nagoya University
5. Chauvet, G.A.: Phil. Trans. R. Soc. Lond. B. 339 (1993) 425-444
6. Jantch, E.: Autopoiesis. A Theory of Living Organization (ed. Zeleny (1981) 65-88 (North Holland, N.Y.)
7. Curie, P.: Journ. de Phys. (III), 3 (1894) 393
8. Koptsik, V.A.: J. Physics vol. C (1983) 16
9. Planck, M.: Z. Phys. 35 (1925) 49-57
10. Planck, M.: Sitzungsber. Acad. Wiss. Berlin (1925) 442-451
11. Eigen, M. & Schuster, P.: The Hypercycle. A Principle of Natural Self-Organization (Springer-Verlag, Berlin) (1979)

Problem Structure and
Fitness Landscapes

An Evolutionary Algorithm for Single Objective Nonlinear Constrained Optimization Problems

Hunter T. Albright and James P. Ignizio

Department of Systems Engineering, University of Virginia, Charlottesville, VA
hta3u@virginia.edu, ignizio@virginia.edu

Abstract. This paper presents research into an evolutionary algorithm that utilizes the *coevolution* of feasible and infeasible solutions in solving constrained optimization problems. The evolution of these populations occurs through the use of traditional and specially designed operators that allow for crossover to occur in each of the populations as well as across the two populations. The cross population crossover allows for the information contained in the infeasible solutions to be utilized in the search for the optimal solution.

1 Introduction

Most real world optimization problems are best represented as constrained optimization problems. The general single objective nonlinear programming problem is formulated as follows:

$$\text{Optimize} \quad f(\mathbf{X})$$
$$\text{Subject to:}$$
$$\mathbf{C}(\mathbf{X}) \ (=, \leq, \geq) \ 0.0$$
$$\mathbf{X} \quad \in \quad \Omega$$

where:

- f is the objective function,
- $\mathbf{X}$ is the vector of decision variables,
- $\mathbf{C}$ is the set of general constraints, and
- Ω is variable space.

The different types of methods that are commonly used to solve nonlinear constrained optimization problems can be grouped into two categories, depending on whether they are based on the use of derivative information. The calculus-based methods require that the objective function and the constraints are twice differentiable, and relies on the calculation or approximation of the Hessian matrix. Such techniques often have difficulty identifying the feasible region [25]. When no restrictions, other than continuous variables, are placed on the constraints, the feasible search space can take any shape. It may be non-convex or not even connected. Consequently, searches for the optimal solution can be extremely difficult and the derivative based techniques quickly fail when the problem lacks linearity, convexity, and differentiablility. Therefore, several techniques have been developed that seek to heuristically solve nonlinear constrained optimization problems without reference to derivative information; genetic algorithms (GAs) [14] are one such technique.

A popular technique in genetic algorithms for handling constraints is the use of a penalty function. The research in this paper presents an initial look at an alternative method, SOCGA (Single Objective Constrained Genetic Algorithm), for handling constraints in single objective nonlinear constrained optimization problems. The method uses a GA with *multiple* populations, which consist, respectively, of feasible and infeasible solutions, and specially designed operators that take advantage of the information provided by the feasible and infeasible solutions. In essence, the technique involves the *coevolution* of two populations. In addition, traditional search techniques have been incorporated into the operators to improve the search process.

This paper is organized as follows. The first section provides an introduction to the material. The second section reviews previous research on genetic algorithms for constrained optimization problems. The third section describes an alternative solution approach to the single objective constrained optimization problem. The forth section presents experimental results. The fifth section contains research conclusions and directions for future research.

2 GAs and Constrained Optimization

Genetic Algorithms(GAs), which were first proposed by Holland [14], are a family of heuristics that are often used to solve unconstrained optimization problems. The theory behind the GA is based on the survival of the fittest. In the canonical GA (see Algorithm 1) there is a population of individuals, in which each individual corresponds to a (normally, feasible) solution. The representation of the solution is often a key to the success of the GA. Operators, such as selection, crossover, and mutation, which emulate the processes in nature, are applied to the current population in order to create the next generation. These processes are applied until the stopping criteria has been met.

Algorithm 1 Canonical Genetic Algorithm

1: $t = 0$
2: generate initial population P_t
3: evaluate P_t
4: rank P_t
5: **while** (not termination-condition) **do**
6:    $t = t + 1$
7:    select parents from P_{t-1}
8:    recombine parents to create P_t
9:    evaluate P_t
10:   rank P_t
11: **end while**

The success of GAs in solving optimization problems can be attributed to several characteristics that are inherent to evolution. Since a population of solutions is maintained, there is implicit parallelism in the search. In addition, the crossover and mutation operators allow for the solution space to be searched in a manner that typically avoids getting stuck in local optimum points. The crossover operator allows for more promising regions to be searched while the mutation operator allows for a more diverse sampling of the solution space.

The handling of constraints remains one of the most challenging aspects of solving optimization problems. For some classes of problems, such as linear programming, elegant algorithms have been devised that take advantage of the characteristics of the constrained search space in order to find the optimal or near optimal solution. However, when the constraints are nonlinear and/or the constrained search space is non-convex, special steps must be taken in order to solve for the solution.

Genetic algorithms have been applied to a large number of unconstrained optimization problems, however their use on constrained optimization problems has not been as extensive. Previous researchers have typically handled constraints in one of three ways:

1. through the use of a penalty function,
2. through a domain specific heuristic, or
3. by restricting the heuristic to some particular objective functions and/or constraints.

In the following sections these techniques are addressed in more detail.

2.1 Penalty Functions

One of the most common approaches to handling constraints is the use of penalty functions. With a penalty function the problem is reformulated as an unconstrained optimization problem,

$$\text{Optimize } f(\mathbf{X}) + \varepsilon \delta \sum_{i=1}^{m} W_i C_i(\mathbf{X})$$
$$\text{Subject to:}$$
$$\mathbf{X} \in \Omega$$

where:

- f is the objective function,
- $\mathbf{X}$ is the vector of decision variables,
- $\mathbf{C}$ is the set of general constraints,
- ε is -1 for a minimization and 1 for a maximization problem,
- δ is a penalty coefficient,
- W_i is a penalty related to the ith constraint,
- m is the number of constraints, and
- Ω is variable space.

Consequently, a wider variety of methods can be applied. This technique has been adopted from the use of penalty functions in calculus based methods for constrained optimization problems. However, some of the problems with penalty functions include the fact that they are domain dependent and that there are not any good guidelines for constructing a general penalty function [17].

Courant [7] is generally attributed as being the first to use penalty functions to solve constrained optimization problems. Camp [4] and Piertgykowski [23] extended the idea to nonlinear constrained optimization problems. However, Fiacco and McCormick [8,9,11,10,12] were the first to make significant progress in applying penalty methods to practical problems. Zangwill [27,28] was also an early contributor to the literature on penalty methods.

The following list represents a survey of methods for GAs which use penalty function to handle constraints.

- The method of Homaifar, Lai, and Qi [1] assumes that for every constraint there is a family of intervals which determine appropriate penalty coefficients.
- The method of Joines and Houch [16] uses dynamic penalties.
- The method of Michalewicz and Attia [18] (Genocop II) first solves the linear constrained optimization problem with Genocop I [19] and then uses a quadratic penalty function to solve the nonlinear constrained part using a standard genetic algorithm.
- The method of Powell and Skolnick [24] uses a variation of the penalty method where infeasible individuals have increased penalties such that the value of the best infeasible solution must be greater than the worst feasible solution.
- The death penalty method is one in which all infeasible individuals are rejected.

For a more comprehensive review of the methods presented above, the reader is directed to Michalewicz [17].

2.2 Domain Specific Methods

Genetic algorithms that use domain specific information often obtain the best results. Such algorithms commonly use the domain specific information to either repair the infeasible individuals or to always generate feasible individuals. The following list represents a survey of domain specific methods.

- The method of Michalewicz [19] (Genocop I) uses a genetic algorithm to solve problems that are linearly constrained.
- The method of Grefenstette uses a genetic algorithm to solve the TSP.
- The method of Paredis [21] is based on constraint propagation for solving job shop scheduling problems.
- The method of Chu and Beasley [6] uses a genetic algorithm to solve set partitioning problems, such as those that might appear for flight crew scheduling.
- The method of Carlson and Shonkwiler [5] uses a genetic algorithm to solve component selection problems.
- The method of Bean and Hadj-Alouane [2] uses a dual genetic algorithm to solve bounded integer programs.
- The method of Hadj-Alouane and Bean [13] uses a genetic algorithm to solve the multiple-choice integer program.
- The method of Brown [3] uses a genetic algorithm to solve constrained facility location problems.

2.3 General Methods

One of the least researched areas in genetic algorithms, outside of penalty function based methods, has been methods for general nonlinear constrained optimization problems. Some published methods include:

- The method of Schoenauer and Xanthakis [25] samples the feasible region by evolving from an initial random population, successively applying a series of different fitness functions which embody constraint satisfaction. The final step is the optimization of the objective function restricted to the feasible region. The success of the whole process is highly dependent on the genetic diversity maintained during the first steps, ensuring a uniform sampling of the feasible region.
- The method of Michalewicz and Attia [18] (Genocop II) first solves the linear constrained optimization problem with Genocop I [19] and then uses a quadratic penalty function to solve the nonlinear constrained part using a standard genetic algorithm.
- The method of Surry, Radcliffe and Boyd (COMOGA) [22] uses a multiple objective genetic algorithm to solve constrained optimization problems.

3 SOCGA

SOCGA is our evolutionary algorithm for constrained single objective optimization problems. The uniqueness of SOCGA is the *coevolution* of a population of *feasible* solutions and a population of *infeasible* solutions in searching for an "acceptable" set of solutions. This set of solutions may include feasible and infeasible solutions depending on the decision maker's tolerance for deviation in the constraints. In Figure 1, a two dimensional solution space is depicted where the shaded area is the feasible region and x^* is the optimal solution. It can be seen that the number of infeasible solutions within a given radius of x^* is much greater than the number of feasible solutions. Consequently, by using the information contained in the infeasible solutions, it is believed that the search for the optimal solution, x^*, is enhanced, as well as, the ability to find "acceptable" solutions which are close to the optimal solution.

Penalty function based methods, the most widely used constraint handling method in GA's, put pressure on the feasibility of a solution through the penalty assigned based on violated constraints. The result is that the search progresses primarily in or near the feasible solution space, see Figure 1. However, with SOCGA, no pressure for feasibility is asserted except for the pressure which naturally occurs through evolution. Therefore, it is possible to make jumps which could, theoretically, advance the search at a much faster rate than a penalty function method would allow.

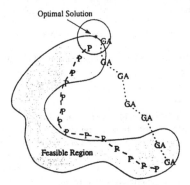

Fig. 1. A possible search path for a penalty function based method is traced out with "P's" and the "GA's" trace out a possible search path for SOCGA. Notice how the search path for a penalty function method primarily stays in or near the feasible region. While SOCGA allows the search to move further away from the feasible region, thus allowing for possible large improvements in the search. Each point in the search paths represent the most fit individual in a GA population for the associated methods.

3.1 Methodology

The first step to utilizing SOCGA in the solution of a constrained optimization problem is the model formulation. A goal programming framework has been used for the model formulation [15]. This framework was selected because the authors believe it to be a good way to model many real world problems and will allow the presented method to be extended to multiple objective problems. The formulation of a general single objective problem in a goal programming framework would be as follows.

Minimize:
$$(\mu^1\rho^1 + \omega^1\eta^1, f(\mathbf{X})) = U$$
Subject to:
$$\mathbf{C}(\mathbf{X}) + \eta_i - \rho_i = 0$$

where:

- U is the achievement vector,
- ρ^1 is the sum of the unwanted positive deviations,
- η^1 is the sum of the unwanted negative deviations,
- μ^1 is a vector of weights associated with the unwanted positive deviations,
- ω^1 is a vector of weights associated with the unwanted negative deviations,

- ρ_i is the positive deviation for the ith constraint,
- η_i is the negative deviation for the ith constraint,
- $f(\mathbf{X})$ is the single objective function,
- $\mathbf{C}(\mathbf{X})$ is the set of constraints, and
- $\mathbf{X}$ is the set of decision variables.

This formulation now allows the different populations to be ranked based on the different priority levels in the achievement vector. The feasible population will be ranked according to the second priority level which corresponds to the objective function. The infeasible population will be ranked according to the first priority level which corresponds to the unwanted deviations of the goals. In both cases we are seeking to minimize the given metric.

Before introducing the main steps of the SOCGA algorithm, the primary characteristics of the method are discussed. These characteristics include the following:

- A floating point representation is used for each chromosome. This representation was selected due to the ease of implementation and the reduction of required memory for large problems.
- The initial pool[1] consists of *two* populations, a population of feasible solutions and a population of infeasible solutions.
- Each solution is evaluated based on two measures, the sum of the *unwanted* deviations from the constraints and the objective function value. These evaluation criteria correspond to the first and second priority levels in the achievement vector of the goal programming model, respectively.
- Crossover was allowed to occur between the feasible and infeasible populations for both the traditional and new operators.
- Elitism was utilized in the creation of the next generation.

In general SOCGA follows the steps of the canonical GA. The pseudo code for the algorithm is shown in Algorithm 2.

Algorithm 2 SOCGA

1: $t = 0$
2: **generate** initial pool P_t
3: **evaluate** initial pool P_t
4: **rank** initial pool P_t
5: **while** (not termination-condition) **do**
6:     $t = t + 1$
7:     **generate** new P_t
8:     **evaluate** P_t
9:     **rank** P_t
10: **end while**

The first task was the generation of the initial pool. While infeasible solutions could readily be generated, filling a population with feasible solutions proved to be much more challenging. This is a common problem and is discussed in more detail in the following section. The second task was the creation of the next generation. This process utilized three different types of operators; elitism, crossover and mutation. Once the new pool had been created, all of the individuals were evaluated using the goal program that had been formulated. The infeasible population was ranked according the to goal deviations where the smallest deviation was preferred and the feasible population was ranked based on the minimization of the objective function.

3.2 Initial Pool

The random generation of a population of feasible solutions can be difficult in a highly constrained problem since the feasible solution space may be small in comparison to the entire solution space. Consequently, it is necessary to take additional steps in order to obtain an initial population of feasible solutions. In Genocop I [19], a set number of random solutions are generated. If a feasible solution is not randomly generated, the user is prompted to enter in a starting feasible solution. The feasible solution is then copied n times in order to generate the initial population. However, this is not a very practical method, since finding a feasible solution can be one of the most difficult problems.

In SOCGA generating the initial pool is begun by generating R random solutions, where R is set by the user. A "Minimal" random number generator of Park and Miller with Bays-Durham shuffle and added

---

[1] Pool is the term used to describe all of the solutions contained in the feasible and infeasible populations.

safeguards was used [26]. If the generated random solution is feasible, it is inserted into the feasible population as long as it is not full and the fitness of the solution is better than the worst member of the population. If the randomly generated solution is infeasible, it is inserted into the infeasible population as long as the population is not full and the fitness of the solution is better than the worst member of the population. Finally, if there is not a full population of feasible solutions, the infeasible population is evolved, where solutions with the smallest deviation are determined to be the most fit, until the feasible population has been filled. By using the deviation as the fitness value and attempting to minimize it, the algorithm is putting an emphasis on the feasibility of a solution. The pseudo code is presented in Algorithm 3.

Algorithm 3 Generate Initial Pool

1: **for** 1 to R **do**
2:    generate a random individual ρ
3:    **if** $\rho \in \mathcal{F}$ **then**
4:      **if** numOfFeasibleSolutions $< n$ **then**
5:        $\rho \rightarrow \mathcal{F}$
6:      **end if**
7:      **if** numOfInfeasibleSolutions $< n$ **then**
8:        $\rho \rightarrow \mathcal{I}$
9:      **end if**
10:    **end if**
11: **end for**
12: **if** numOfFeasibleSolutions $= 0$ **then**
13:    evolve infeasible population until X feasible solutions exist
14: **end if**
15: **if** numOfFeasibleSolutions $< n$ **then**
16:    **for** 1 to populationSize - numOfFeasibleSolutions **do**
17:      randomly select one of the feasible solutions
18:      copy the feasible solution into the feasible population
19:    **end for**
20: **end if**

After the initial pool has been generated, the solutions in each of the populations are evaluated based on two measures of fitness, the value of the objective function and the sum of the unwanted deviations from the constraints. The populations are then ranked *lexicographically* based on the two measures. In the feasible population, the objective function value has the highest priority. In the infeasible population, the sum of the unwanted deviations from the constraints has the highest priority.

3.3 Next Generation

The creation of the next generation is controlled through the use of several different operators. The creation of the next generation is effectively a search through the solution space. This search is directed through the use of different operators. In SOCGA both traditional and specially designed operators were used to enhance the search. These operators included the following.

Selection The selection operator is used to select parents from the current generation for the next generation. The parents are selected based on their fitness. A linear selection operator was used which selected parents from the most fit individual to the least fit individual with probabilities that decreased in a linear fashion.

Elitism The elitism operator helps to preserve the best solutions from one generation to the next by allowing a selected number of the best individuals to survive.

Crossover The crossover operator is largely responsible for fine tuning the search in regions of interest through the recombination of selected individuals. Both traditional and specialized crossover operators were used. The traditional operators used include uniform and two point crossover methods.

Mutation The mutation operator helps ensure that the search does not get stuck in a local optimum and the search is sufficiently diverse. This is achieved through the random mutation of selected genes in given individuals.

One of the key characteristics of the crossover operators is that they allow for crossover to occur in both of the populations and *across the two populations.* By allowing crossover to occur between the feasible and infeasible populations, the information that is contained in the infeasible solutions is incorporated in the search for the optimal solution. The specially designed operator is referred to as a *genetic flip*, which is based on a simplex search, in which the least preferred point of the simplex is reflected through or contracted towards the base of the simplex.

The genetic flip operator is based on the downhill simplex method of Nelder and Mead [20]. A simplex is a geometrical figure consisting, in N dimensions, of N + 1 points and all interconnecting lines and faces. The method requires only function evaluations and allows for the use of feasible and infeasible chromosomes. A description of how the genetic flip is implemented is shown below.

Genetic Flip (for a 2-dimensional problem) produces a single child from N + 1 parents in N dimensions. This is operator represents a traditional technique in optimization, yet it is believed to have promise as an operator in the proposed method due to the ability to work in the entire solution space. The child is created through the following steps (see Figure 3):

1: Select point $p_i \in (\mathcal{F}, \mathcal{I})$ where $i = 1, ..., N + 1$.
2: Evaluate $f(p_i)$.
3: Rank p_i based on fitness.
4: Reflect(simplex) $\rightarrow p_{new}$.
5: **if** $f(p_{high}) \succ^2 f(p_{new})$ **then**
6:     contract(simplex) $\rightarrow p_{new}$.
7: **end if**

3.4   Example Results

This section illustrates the solution method through a simple two variable example. Suppose the problem, $EX1$, is:

Lexicographically Minimize:
$$(\rho_1 + \eta_2, -x_1 - x_2) = U$$
Subject to:
$$x_1^2 - 10 - x_2 + \eta_1 - \rho_1 = 0$$
$$-x_1^2 + 10 - x_2 + \eta_2 - \rho_2 = 0$$

where:

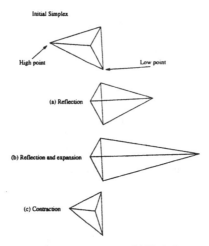

Initial Simplex

High point   Low point

(a) Reflection

(b) Reflection and expansion

(c) Contraction

Fig. 2. With a given simplex there are three basic moves. (a) The high point can be reflected through the base of the simplex. (b) The high point can be reflected through the base of the simplex and the simplex can be expanded in one or more directions. (c) The simplex can be contracted in one or more directions. In each of these moves it is assumed that the low point is preferred to the high point.

---

2 The symbol $\succ$ means "is preferred to".

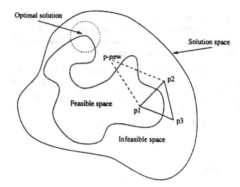

Fig. 3. Three solutions, p_1, p_2, and p_3, are selected to anchor the simplex. A new solution, p_{new}, is generated by executing a legal move with the simplex.

- U is the achievement vector,
- ρ_i is the positive deviation for the ith constraint,
- η_i is the negative deviation for the ith constraint,
- x_1 and x_2 are the decision variables, and
- the underlined variables represent the unwanted deviations.

The two equations are the two constraints. The single objective of the problem is to maximize $(x_1 + x_2)$. However, the problem has been cast as a lexicographic minimization problem in order to work with the proposed method.

generations	5
population	10
constraints	2
chromosome length	2
crossover rate	1
mutation rate	0.5
elitism crossover mutation elitism	yes
variable max	10 10
variable min	0 0

Table 1. GA Parameter Settings for EX1

The example problem was run with and without cross population crossover. The distribution of the final feasible solutions is shown in Figure 4. The feasible solutions that resulted from the GA in which cross population crossover was allowed are indicated by the small diamond shape, ⋄, in the figure. The data indicated by a small plus, +, is the final population of feasible solutions in a GA where cross population crossover was not allowed. The solutions when cross population cross over was allowed have a greater variance, but are closer to the optimal solution (0.5, 9.75). The parameters settings for the example problem are shown in Table 1.

With Co-Evolution		Without Co-Evolution	
FpFpFc 22	FpFpIc 0	FpFpFc 69	FpFpIc 1
IpIpFc 3	IpIpIc 13	IpIpFc 0	IpIpIc 0
FpIpFc 25	FpIpIc 15	FpIpFc 0	FpIpIc 0
Totals 50	28	69	1

Table 2. Results of Parent Combinations

Another interesting result was the distribution of feasible and infeasible children resulting from different parent combinations. In Table 2, it can be seen that the number of feasible children that resulted from crossover with infeasible parents comprises 56% of the feasible children. *These results are an indication of the potential value the infeasible solutions can have in the solution process.*

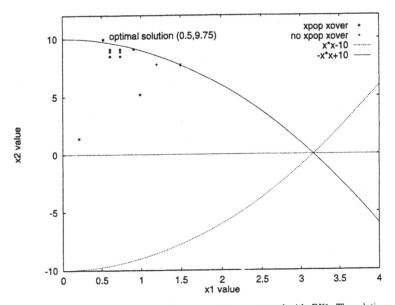

Fig. 4. Two different strategies for handling constraints were tested with *EX*1. The solutions indicated by the small diamond shape, ⋄, are the feasible solutions that were produced from a GA in which cross population crossover was allowed. The solutions indicated by a small plus, +, are the solutions that were produced from a GA in which cross population crossover was not allowed. It can be seen that the solutions produced by a GA utilizing cross-population crossover are closer to the optimal solution of (0.5, 9.75).

4 Conclusions

SOCGA is a heuristic algorithm that utilizes the information in infeasible solutions to improve the search process for a set of "acceptable" solutions for a nonlinear constrained optimization problem. This is achieved through the *coevolution* of a population of feasible solutions and a population of infeasible solutions. The evolution of these populations occurs through the use of traditional and specially designed operators that allow for crossover to occur in each of the populations as well as across the two populations. The cross population crossover allows for the information contained in the infeasible solutions to be utilized in the search for the optimal solution.

The initial results are promising. They show that when the information contained in the infeasible solutions is used, the search for "acceptable" solutions is improved and a greater number of solutions are found which fit the "acceptable" description. Areas for future research include the following:

– Additional empirical tests to access the true value of coevolving feasible and infeasible populations.
– Experimentation on a set of problems comparing SOCGA with a penalty based GA.

References

1. S.H.-Y. Lai A. Homaifar and X. Qi. Constrained optimization via genetic algorithms. *Simulation*, 62:242–254, 1994.
2. James C. Bean and Atidel Hadj-Alouane. A dual genetic algorithm for bounded integer programs. Technical report, Department of Industrial and Operations Engineering, The University of Michigan, 1992.
3. E. Brown. *Using the Facility Location Problem to Explore Operator Policies and Constraint-Handling Methods for Genetic Algorithms*. PhD thesis, University of Virginia, Charlottesville, VA, April 1996.
4. G.D. Camp. Inequality-constrained stationary-value problems. *Operations Research*, 3:548–550, 1955.
5. Susan E. Carlson and R. Shonkwiler. Annealing a genetic algorithm over constraints. 1995.
6. P.C. Chu and J.E. Beasley. A genetic algorithm for the set partitioning problem. 1995.
7. R. Courant. Calculus of variations and supplementary notes and exercises. Technical report, New York University, 1962.
8. A.V. Fiacco and G.P. McCormick. Computational algorithm for the sequential unconstrained minimization technique for nonlinear programming. *Management Science*, 10:601–617, 1964.

9. A.V. Fiacco and G.P. McCormick. Extentions of sumt for nonlinear programming: Equality constraints and extrapolation. *Management Science*, 12:816–828, 1966.

10. A.V. Fiacco and G.P. McCormick. The sequential unconstrained minimization technique (sumt), without parameters. *Operations Research*, 15:820–827, 1967.

11. A.V. Fiacco and G.P. McCormick. The slacked unconstrained minimization technique for convex programming. *SIAM J. Applied Mathmematics*, 15:505–515, 1967.

12. A.V. Fiacco and G.P. McCormick. *Nonlinear Programming: Sequential Unconstrained Minimization Techniques*. John Wiley and Sons, New York, 1968.

13. Atidel Hadj-Alouane and James C. Bean. A genetic algorithm for the multiple-choice integer program. Technical report, Department of Industrial and Operations Engineering, The University of Michigan, 1993.

14. John Holland. *Adaptation in Natural and Artificial Systems*. University of Michigan Press, Ann Arbor, Michigan, 1975.

15. J.P. Ignizio. *Introduction to Linear Goal Programming*. Sage Publishing, Beverly Hills, CA, 1985.

16. J.A. Joines and C.R. Houch. On the use of non-stationary penalty functions to solve nonlinear constrained optimization problems with gas. In *Proceedings of the Evolutionary Computation Conference - Poster Session*, pages 579–584, Orlando, FL, June 1994. IEEE World Congress on Computational Intelligence.

17. Zbigniew Michalewicz. A survey of constraint handling techniques in evolutionary computation methods. *Found on the internet*, 1995.

18. Zbigniew Michalewicz and N. Attia. Evolutionary optimization of constrained problems. In A.V. Sebald and L.J. Fogel, editors, *Proceedings of the 3rd Annual Conference on Evolutionary Programming*, pages 98–109, River Edge, NJ, 1994. World Scientific Publishing.

19. Zbigniew Michalewicz and Cezary Janikow. Handling constraints in genetic algorithms. In R.K. Belew and L.B. Booker, editors, *Proceedings of the Fourth International Conference on Genetic Algorithms*, pages 151–157, San Mateo, CA, 1991. Morgan Kaufmann.

20. J.A. Nelder and R. Mead. *Computer Journal*, 7:308–313, 1965.

21. J. Paredis. Co-evolutionary constraint satisfaction. In *Proceedings of the 3rd Conference on Parallel Problem Solving from Nature*, pages 46–55, New York, 1994. Springer-Verlag.

22. Nicholas J. Radcliffe Patrick D. Surry and Ian D. Boyd. A multi-objective approach to constrained optimization of gas supply networks: The comoga method. 1995.

23. T. Pietrgykowski. Application of the steepest descent method to concave programming. In *Proceedings of International Federation of Information Processing Societies Congress*, pages 185–189, Amsterdam, 1962. North-Holland.

24. D. Powell and M.M. Skolnick. Using genetic algorithms in engineering design optimization with non-linear constraints. In *Proceedings of the Fifth International Conference on Genetic Algorithms*, pages 424–430, Los Altos, CA, 1993. Morgan Kaufmann Publishers.

25. Marc Schoenauer and Spyros Xanthakis. Constrained ga optimization. In S. Forrest, editor, *Proceedings of the Fifth International Conference on Genetic Algorithms*, pages 573–580, Los Altos, CA, 1993. Morgan Kaufmann.

26. William Vetterling William Press, Saul Teukolsky and Brain Flannery. *Numerical Recipes in C: The Art of Scientific Computing*. Cambridge University Press, New York, second edition, 1994.

27. W.I. Zangwill. Nonlinear programming via penalty functions. *Management Science*, 13:344–358, 1967.

28. W.I. Zangwill. *Nonlinear Programming: A Unified Approach*. Prentice-Hall, Englewood Cliffs, N.J., 1969.

On Recombinative Sampling

Ian R. East

57, Kidlington Road, Islip, Oxfordshire OX5 2SS, England
ian.east@dial.pipex.com

Abstract

If a population is constrained to exhibit no variation in the allelic frequency distribution of any gene then information may only be recorded by recombining genes into new configurations (schemata). It is shown that such a constraint need lead to no loss of information capacity. A simple algorithm, employing direct replacement and a single uniform genetic operator, is then analysed with regard to schema sampling. The probabilities of schema creation and destruction are proven identical, regardless of operator. The probabilities are then deduced, with which new schemata are sampled by, first, recombination, and, second, mutation, in relation to order and allelic variation. The analysis overcomes limitations inherent in earlier work [4, 9, 10] and yields simple expressions, computationally affordable for any schema order, and valid for any size of allelic alphabet. Recombination is shown capable of matching any sampling rate mutation can offer, and of exploiting a decline in admitted allelic variation.

1 Introduction

1.1 Aims

The work reported here concentrates on understanding the effects of recombination. To this end, a simplified genetic algorithm is considered, where recombination alone acts to modify a randomly initialised population. Direct replacement is assumed in order to maintain a constant flat allelic distribution in every gene. Such a simple algorithm has been found to be effective [3, 5].

Holland's Schema Theorem predicts how schemata, that exist within the population, will prosper. Here, we seek to determine the rate at which recombination samples *new* schemata, and compare it with that possible via mutation. We also seek the variation in schema sampling rate with schema order, and the number of available alleles of each gene.

1.2 Background

Much effort has been expended in the attempt to understand the *mechanism* by which a genetic algorithm is able to continuously increase the fitness of its population. The Schema Theorem [7] guarantees that biased selection and replacement will cause existing schemata to prosper in accordance with their perceived utility. It says nothing about the sampling of *new* schemata. Recombination has long been regarded merely as a threat to schema proliferation. Early analysis aimed to deduce the probability of schema "disruption" [4]. Recently it has been appreciated that recombination is also engaged in schema creation.

De Jong's early analysis took no account of the possibility that parents might share the same allele at a position defined by a given schema. This was later corrected by Spears and De Jong [9]. The correction makes a profound difference, both to the predicted disruption probability, and to the dependence of that probability on schema defining length. Their analysis results in an expression that is computationally unaffordable for all but the smallest schema order. The reason for this is that they integrate over all possible distributions of defined positions in order to produce an expectation value. This is only reasonable if all distributions are equally likely. For this to be so, no selection bias must act. The analysis is also limited to binary representation.

Spears and De Jong do address the issue of schema sampling by recombination, which they term "productivity". They note that the analysis of schema disruption is identical to that of schema production. This does not infer the equality of the probabilities of creation and destruction of the *same*, arbitrarily chosen, schema. When one schema is destroyed, another is created. However, the schema destroyed and the one created are both distinct and correlated.

More cross-points are shown to increase the probability of schema disruption and reduce the influence of defining length. It is possible to demonstrate that such an increase is dependent upon cross-point *distinguishability*, which yields a combinatorial advantage. In nature, cross-points are indistinguishable. Increasing their number thus decreases schema disruption probability, and hence productivity. Nature employs few cross-points.

Spears later employed the analysis to compare mutation and recombination with regard to schema productivity [10]. Recombination productivity had already been noted to diminish with falling population diversity. Exploration is increasingly confined according to the remaining allelic diversity. Mutation productivity remains constant. He claims that recombination affords an advantage over mutation by simultaneously offering a high probability of both construction and survival. This is shown impossible below.

Spears' productivity comparison is based upon specific examples where each operator constructs a given new schema from a given existing one. The comparison remains interesting and informative but is insufficient as a basis for a quantitative comparison of sampling capability. Account is taken of falling allelic diversity via the probability of allelic matching. This is adequate only when there is either no linkage or no selection bias.

Geiringer derived recurrence relations governing the frequency of any given genotype, or Holland schema, under just recombination and random selection [6]. These are valid for any degree of allelic variation but would prove prohibitively expensive in practice. Booker employed her analysis to deduce schema transmission probabilities [1]. His results only cater for binary genes, and are misleading because no account is taken of occasions when both parents share defined alleles.

Because the term "construction" is strongly associated with the Building-Block Hypothesis, it is here abandoned. This paper presents a view whereby the sampling of schemata is driven by recombination, and biased via selection. No such hypothesis is necessary. Also, "disruption" seems now to infer a wholly undesirable event. It is not so. Each time it occurs, some new schema is born. The terms "creation" and "destruction" are preferred.

2 Population as combinatorial memory

In an algorithm where offspring directly replace parents, information may only be stored by genetic recombination and not by varying the allelic distribution within any gene independently. It might at first seem that the loss of the latter will seriously reduce the capacity of the population memory. This need not be so.

Shannon [8] showed that a natural definition of the information H communicated per symbol, chosen from m options, each with prior probability p_i, is as follows.

$$H = -\sum_{i=1}^{m} p_i \log_2 p_i$$

Here, the "information" of interest is represented by increasing the joint probability of two or more symbols occurring together. Dependence between symbols *decreases* information according to the Shannon definition. Shannon's definition is in fact the population entropy.

Imagine the population arranged in two dimensions, with each member drawn horizontally. The population may be regarded as a sequence of vertical slices, which are initially independent. The entropy of a population of size n, and where there exist two alleles of every gene, will be given by the following.

$$H_1 = l . \log_2 \binom{n}{\frac{n}{2}}$$

This is the maximum entropy possible. Equal numbers of each allele in every gene maximizes the number of possible population states. Random instantiation gives each state equal probability. No genetic information is then recorded. The maximum information has been recorded when a single slice allows the determination of all others. It has then become impossible to record any other genetic combination. The entropy is then as follows.

$$H_2 = \log_2 \binom{n}{\frac{n}{2}}$$

The genetic information capacity of the population is given by the difference between the two entropies. Using Stirling's approximation, it is possible to deduce the following.

$$I = H_1 - H_2 = (l-1).\log_2 \binom{n}{\frac{n}{2}} \approx l.n \qquad\qquad (l \gg 1, n \gg 1)$$

Provided both population size and genome length are both significantly greater than unity, constraining the population to exhibit uniform allelic distributions does not reduce its capacity for genetic information.

3 Recombinative sampling

3.1 Schema creation and destruction

As noted above, the probabilities of creation and destruction of the same arbitrary schema have *not* so far been proven identical. The proof is trivial but has not been previously reported.

A "uniform" genetic operator is one which alters each gene with equal probability, and in an identical fashion. All crossover recombination conforms to this definition.

The theorem may be stated as follows, for an arbitrary uniform genetic operator x.

$$\forall \xi. p_{cx}(\xi) = p_{dx}(\xi) \qquad p_{cx}(\xi) = \textit{probability of creation of schema } \xi.$$
$$p_{dx}(\xi) = \textit{probability of destruction of schema } \xi.$$

The following axiom would not appear contraversial.

➤ Under the influence of any single uniform genetic operator, and in the absence of selection bias, the same probability of creation will be shared by all schemata that compete within the same defining positions.

$$\forall \xi \forall \xi'. p_{cx}(\xi) = p_{cx}(\xi') \qquad\qquad \xi, \xi' \textit{ define same positions.}$$

Choose an operator and a schema, both at random. Assume the theorem false. Now suppose that the operator has acted and destroyed the schema. Whatever the operator, a new schema will have been simultaneously created, within one offspring, with the same positions defined as the original. Thus the probability of creation of the new schema equals that of the destruction of the original. If the theorem is false then the probabilities of creation of each schema need not equate. The operator may thus favour one above the other, in conflict with the axiom above.

3.2 Schema sampling by 1-point crossover recombination

To obtain schema creation probability we borrow a method previously employed to compute the probability of constructing an entire genotype [2]. A member of the population may only be considered a potential parent if its genotype exhibits a schema that subsumes[1] the chosen one. Parental subsuming schemata must juxtapose or overlap for success to be possible. This approach is applicable to n-point crossover, for any value of n. On this occasion, we shall only consider 1-point crossover.

o = *order of schema to be constructed.*

d = *defining length of schema to be constructed.*

σ = *order of overlap (correctly defined positions in both parent genotypes).*

$\delta(\sigma)$ = *defining length of overlap.*

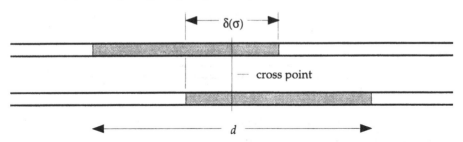

Figure 1 Criterion for a chosen schema to be created by 1-point crossover recombination.

If there exists an overlap between subsuming schemata then the cross-point must fall within it. If they juxtapose then it must fall between them. An overlap becomes less likely with increasing size, but it then becomes more likely that the cross-point will fall within it. The probability of creation is the product of the probability of a given overlap and that of a recombination which then successfully creates the desired schema. The overlap probability may be expressed as the proportion of possible parent genotypes which offer it.

Because it will be necessary to sum probabilities of all possible overlaps, care must be taken not to count the same configuration more than once. To this end we begin with zero overlap order (i.e. juxtaposition) and count upwards.

---

[1] Any schema defines a set of genotypes. One schema is said to *subsume* another if the set it defines includes that defined by the other. The lower the schema order the larger the set it defines.

It is then only necessary to ensure that each subsuming schema is punctuated with an allele which conflicts with the desired schema in order to ensure each configuration is only accounted for once. An example is shown in Figure 2 below.

Each configuration is regarded as a single outcome of selection, not as two distinct outcomes as would be the case if parents were distinguishable. Configurations (outcomes of selection) must be counted accordingly. The alphabet of each gene is assumed to be of equal size. With direct replacement, either offspring may contribute the chosen schema.

$$p_{cr}^{\sigma} = p_{conf}^{\sigma} \times p_{cross}^{\sigma} = \frac{1}{k^{o}(l-1)}\left(\frac{k-1}{k}\right)^{2}\frac{(\delta(\sigma)+1)(o-(\sigma+1))}{k^{\sigma}}$$

In order to determine schema creation probability, regardless of the order of overlap, it is necessary to sum p_{cr}^{σ} over σ, from 0 to o - 2. Before this can be done, some assumption regarding the distribution of defined positions across the genome must be made. Varying the distribution of defined positions has little effect upon creation probability while defining length remains constant. (It is possible to modify the analysis to verify this.)

☑ Gene matching that of chosen schema   ☒ Gene that does *not* match

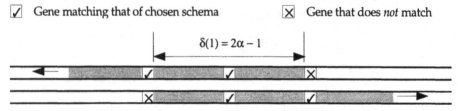

Figure 2 The dependency of length upon order for schema overlap.

Consider schemata whose defined positions are evenly distributed along the genome. These may now be regarded as *representative* of all schemata. The liberty remains to vary the schema defining length via the interval α between defined positions.

$$\delta(\sigma) = (\sigma+1).\alpha - 1 \qquad \alpha = \textit{interval between defined positions.}$$

$$\alpha = \psi.\frac{l+1}{o+1} \qquad\qquad \frac{o+1}{l+1} \leq \psi \leq \frac{o+1}{o}$$

$$\Rightarrow \quad p_{cross}^{\prime\sigma} = \alpha\frac{\sigma+1}{l-1} = \psi\frac{l+1}{o+1}\frac{\sigma+1}{l-1} \approx \psi\frac{\sigma+1}{o+1} \qquad (l >> 1)$$

$$\Rightarrow \quad p_{cr}^{\prime\sigma} \approx \psi\frac{1}{(o+1)k^{o}}\left(\frac{k-1}{k}\right)^{2}\left(\frac{(o-1)+(o-2)\sigma-\sigma^{2}}{k^{\sigma}}\right) \qquad (l >> 1)$$

An expression may now be deduced for the probability $p_{cr}^{\prime}$ of constructing a schema, of the kind described above, in terms of only schema order and allelic variation. Only a very mild restriction regarding genome length has proved necessary. Neither o nor k are constrained.

$$p'_{cr} = \sum_{\sigma=0}^{o-2} p'^{\sigma}_{cr} \approx \psi \frac{1}{(o+1)k^o}\left(o\left(1+\frac{1}{k^o}\right) - \frac{k+1}{k-1}\left(1-\frac{1}{k^o}\right)\right) \qquad (k>1, l \gg 1)$$

For schemata whose order significantly exceeds unity the expression reduces. Also, as the ratio of schema order and genome length grows, the range of, and distribution over, ψ changes in favour of higher values.

$$p'_{cr} \approx \psi \frac{1}{k^o} \rightarrow \frac{1}{k^o} \qquad (l \gg 1, o \gg 1, o \rightarrow l)$$

3.3 Schema sampling by homogeneous mutation

By "homogeneous mutation" we mean that each gene is considered independently and equally, regardless of its position. With probability p_m, a new allele is assigned, chosen from the appropriate alphabet. All alleles in the alphabet share the same selection probability. Note the possibility that a mutation might yield no change. The alphabet of every gene is assumed of equal size k.

Suppose that the recombination operator is replaced with homogeneous mutation. Offspring will no longer be constrained to the allelic content of their parents. However, if both mutation and selection are unbiased then allelic diversity must remain complete and unvarying with time. Homogeneous mutation is a uniform genetic operator. Therefore, probabilities of schema creation p_{cm} and destruction p_{dm} will again be identical.

The probability p_{cm} of creating any new arbitrary schema of order o is simply the product of the probability of randomly selecting a suitable "parent" and that of its successful mutation. A suitable parent is any with genotype exhibiting v positions that mismatch the desired schema, such that v lies between 1 and o. A successful genotype mutation is then one that leaves matching genes unchanged, and modifies the remainder correctly.

$$p^v_{cm} = \binom{o}{v}\left(\frac{1}{k}\right)^{o-v}\left(\frac{k-1}{k}\right)^v\left(p_m\frac{1}{k}\right)^v\left(1-p_m+p_m\cdot\frac{1}{k}\right)^{o-v}$$

$$\Rightarrow \quad p^v_{cm} = \frac{1}{k^o}(1-p'_m)^o\binom{o}{v}\left(\frac{p'_m}{1-p'_m}\right)^v \qquad\qquad p'_m = \frac{k-1}{k}p_m$$

p'_{cm} represents the probability of a mutation that causes change. When $k = 2$ it has a maximum of 0.5, but approaches p_m as k increases.

With a little help from the Binomial Theorem, it is now possible to acquire the probability that a single parent will be selected and mutated to yield a new schema instance.

$$p_{cm} = \frac{1}{k^o}(1-p'_m)^o\sum_{v=1}^{o}\binom{o}{v}\left(\frac{p'_m}{1-p'_m}\right)^v = \frac{1}{k^o}\left[1-(1-p'_m)^o\right]$$

The purpose of this exercise is to allow comparison between the rates at which recombination and mutation sample new schemata. To afford a true comparison we must remember that the algorithm passes the genetic operator *two* individuals, whose offspring are evaluated and then optionally returned to the population. The correct frequency (new instances of chosen schema per operation), for comparison, is as follows.

$$f_{cm} = 2p_{cm}$$

4 Conclusions

4.1 Recombinative sampling

It has been easily shown that the probabilities of creation and destruction (p_{cr} and p_{dr}) of any arbitrarily chosen schema must be exactly equal for *any* uniform genetic operator. Varying either the number of distinguishable cross-points, or the exchange probability given uniform crossover, thus simultaneously dictates both the rate at which new schemata are sampled and their lifetime. An operator that samples new schemata at a greater rate than another will also destroy them more rapidly.

Spears [10] suggested that recombination possesses an advantage over mutation in that it simultaneously achieves a high probability of both schema creation and survival. No such advantage is possible for any uniform genetic operator, including recombination.

In an algorithm where recombination is the only active genetic operator, the sampling rate has been quantified for schemata with evenly distributed defined positions. The creation probability obtained is independent of genome length, and scales linearly with defining length between stated bounds. The bounds upon creation probability thus determined also apply to schemata with an uneven distribution of defined positions. The expression derived offers cheap and simple computation for any order and permitted allelic variation.

A proper basis has been established for comparison between sampling by mutation and sampling by recombination. Simple 1-point crossover recombination is clearly able alone to maintain a sampling rate comparable with any that mutation can offer. The sampling rate afforded by mutation approaches that of recombination exponentially both with effective mutation rate and with schema order.

4.2 Genetic representation

It is apparent from Figure 3 that the rate of schema sampling by recombination is significantly reduced by increasing representation cardinality. It is a maximum with binary genes. However, care is required in the interpretation of this observation.

Probability of Creation

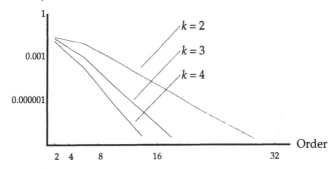

Figure 3 Probability of schema creation by 1-point crossover recombination versus order and allelic variation.

While binary representation clearly offers the highest schema sampling rate, the size of the domain partition dictated by a Holland schema also varies as $1/k^o$, in approximate accord. Any partition, of a given size, may be represented by a schema of lower order with higher k, and will thus be sampled at the same rate. Recombinative sampling of any domain occurs at a rate that is almost independent of the number of its dimensions. The only apparent advantage afforded by reducing representation cardinality is to avoid the reduced sampling rate that would otherwise accompany very low schema order.

4.3 Population convergence

In the algorithm under investigation, allelic diversity within the population is constrained to remain constant. However, selective discrimination would gradually diminish the expected number of alleles of each gene available for recombination. Simply reducing k affords an idealized, but instructive, model. We can now ask how the recombinative schema sampling rate might vary as the number of alleles competing for each defined position reduces towards unity. $p'_{cr}(o,k)$ affords an appropriate measure.

Figure 4 shows that recombination offers exactly what is most desirable. As k falls, recombination is ever more likely to create any schema chosen from those which remain possible. The increase in creation probability is ever more marked, the bigger the schema chosen. Only when the absurd situation is closely approached, where only one allele exists for any gene, does the curve fall back to zero. This is because it becomes impossible for a *new* instance of the desired schema to be created. It becomes much too likely that one or both parents will already safely carry one, denying the chance for another to be born.

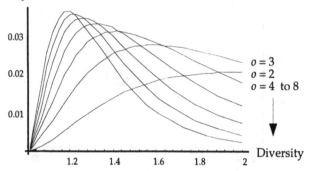

Figure 4 Probability of schema creation by 1-point crossover recombination versus uniform allelic diversity.

References

1. Booker, L. B.: 1992, "Recombination Distributions for Genetic Algorithms", in "Proceedings of The Second Workshop on the Foundations of Genetic Algorithms", Whitley, L. D. (Ed.), Morgan Kaufmann, pp. 29-44.

2. Bridges, C. L. and D. E. Goldberg: 1987, *"An Analysis of Reproduction and Crossover in a Binary-Coded Genetic Algorithm"*, in "Proceedings of Second International Conference on Genetic Algorithms", Grefenstette, J. J. (Ed.), Lawrence Erlbaum Associates.

3. Culberson, J. C.: 1993, *"Crossover versus Mutation: Fueling the Debate: TGA versus GIGA"*, in "Proceedings of the Fifth International Conference on Genetic Algorithms", Forrest, S. (Ed.), Morgan Kaufmann, p. 632.

4. DeJong, K. A.: 1975, *"An Analysis of the Behaviour of a Class of Genetic Adaptive Systems"*, Doctoral thesis, University of Michigan, Department of Computer and Communications Sciences, University of Michigan, Ann Arbor, Michigan, USA.

5. East, I. R., & J. Rowe: 1994, *"Direct Replacement: A Genetic Algorithm without Mutation Avoids Deception"*, in *"Progress in Evolutionary Computation: AI '93 and AI '94 Workshops on Evolutionary Computation"* Lecture Notes in Artificial Intelligence Vol. 956, Xin Yao (Ed.), Springer-Verlag, pp. 41-48.

6. Geiringer, H.: 1944, *"On the probability theory of linkage in Mendelian heredity"*, Annals of Mathematical Statistics, 15, pp. 25-57.

7. Holland, J. H.: 1975, *"Adaptation in Natural and Artificial Systems"*, University of Michigan Press.

8. Shannon, C. E. and W. Weaver: 1949, *"The Mathematical Theory of Communication"*, (1963) Edition, University of Illinois Press.

9. Spears, W. M. and K. A. D. Jong: 1990, *"An Analysis of Multi-Point Crossover"*, in "Proceedings of the First Workshop on the Foundations of Genetic Algorithms", Rawlins, G. J. E. (Ed.), Morgan Kaufmann, pp. 301-315.

10. Spears, W. M.: 1992, *"Crossover or Mutation?"*, in "Proceedings of The Second Workshop on the Foundations of Genetic Algorithms", Whitley, L. D. (Ed.), Morgan Kaufmann, pp. 221-237.

The Evolution of Mutation, Plasticity and Culture in Cyclically Changing Environments

Tony Hirst

Dept. of Psychology, Open University, Milton Keynes, MK7 6AA, UK

Abstract. In this paper, I describe an experiment in which an evolving population is set the task of tracking an environmental value that cycles sinusoidally over several generations. The evolution of individual mutation rates is investigated for a range of cycle lengths. The effect on the evolutionary dynamic of plasticity in the form of simple learning, itself subject to evolution, is investigated, under conditions of both simple inheritance and the inheritance of acquired characteristics (IAC). Finally, the coevolution of mutation and plasticity rates is considered, and the interaction of plasticity and IAC described.

1 Introduction

Over the last decade, genetic algorithms (GA's) have been deployed with some success over a range of stationary optimisation problems. On well defined, temporally varying optimisation functions, however, there has been little work done. Where the environment is non-stationary, the use of convergence around an individual to identify when an optimum has been found is not appropriate, since individuals are assumed to be specialists rather than generalists, and so the population can never settle on a single individual that is optimal over a period of time.

In this paper, I will use an evaluation function that originally owes to Cobb [1], and was also used by Dasgupta [2] as an initial testbed for his structured genetic algorithm. In its original formulation, the evaluation function optimum changed *between* generations and was given by mimimising the error between an individual, p, and a time varying optimum, h_t:

$$h_t = 1.0 + \sin(a \times G\,eneration) \tag{1a}$$

$$e_t(p) = (p - h_t)^2 \tag{1b}$$

$$f_t(p) = -\log_{10} e_t \tag{1c}$$

where p is a 32 bit Gray coded phenotypic individual over the range [0.0, 2.0], derived from 32 genotypic 'targeting' bits (as opposed to 'rate' bits which specify individual mutation rates, for example); h_t gives an environmental state that varies sinusoidally over time, changing between generations, with rate parameter, a, and period $2pi/a$; the individual error is given by e_t and f_t is the fitness value (thus defining a *maximisation* problem).

Previously reported results using this evaluation function [1, 2] have concentrated on the ability of the *fittest* population member at any one time (i.e. the one with the lowest evaluation) to track the environmental state. Taking a cue from the theoretical/quantitative genetics approach of Lande [3], and as discussed in [4], I will focus on the behaviour of population means, although best of generation results are also reported in line with the previous work.

Cobb identified two strategies for coping with fluctuating environments - memory (e.g. through diploidy) and introducing variation (at the *genotypic* level through *mutation*). In contrast, I consider introducing *phenotypic* variation through plasticity.

The effect of three adaptive strategies investigated in *stationary* environments for evaluation landscapes of various degrees of ruggedness in [5] will be considered:

> an individually evolved mutation rate, that sets the bit rate at which an individual is mutated at the start of a generation (rate loci are not subject to the evolved rate);
>
> individual learning at an individually evolved 'learning' rate (this loosely corresponds to the *bias* of Turney [6]);
>
> 'cultural inheritance' or the inheritance of characters acquired through learning.

2 Tracking Fluctuating Environments with Mutation

In Cobb's original experiment, she compared the effectiveness of fixed mutation rates and an adaptive mutation rate strategy on the basis on offline (time averaged generational best) fitness. The adaptive mutation strategy switched from a low to a high globally applied mutation rate whenever the offline fitness dropped for a range of environmental velocities. For comparison, I shall use a simple GA with fixed and evolving mutation rates (m-rates (units bit^{-1})) and will report offline and online (time averaged population mean) fitnesses, although ongoing work ([5]) suggests that these are not necessarily the most appropriate measures, especially when 'settling time' (initial exploration) and equilibrium (steady state) modes are separated out. The individual, evolvable rate is 7 bit binary coded value over [0.0, 1.0], and is used in the biased flipping of each targeting bit following crossover (i.e. the evolved mutation rate is not applied to the bits that code for the rate itself). In addition, mutation at the low background rate is applied equally to all bits in the genome (i.e. targeting and evolved rate bits alike).

All experiments were carried out using a modified version of Genesis 5.0. Unless otherwise specified: graphed results represent the mean of 10 runs, with x-axis representing generation number; population size was 200; the background mutation rate was set at 0.001/bit in all cases; two point crossover was applied at a rate of 0.6 to individuals selected using linear ranking selection with rank minimum 0.5.

2.1 To Compare a Simple GA with Fixed and Evolvable Mutation Rates

In this first experiment, I shall concentrate on Cobb's original results and compare them to results obtained from an algorithm employing individual, evolved mutation rates. Regime GA-M corresponds to the evolved m-rate case; SGA corresponds to a simple genetic algorithm, with fixed m-rates, set at the 'optimal' mutation rates suggested by Cobb as follows -(a:m-rate): 0.001:0.01, 0.01:0.05, 0.05:0.1, 0.1:0.5, 0.5:0.5. These correspond to the generational periods 6283, 628, 126, 63 and 13 respectively.

Figure 1a shows how the optimal fixed rate (SGA) offers better offline performance than the evolved rate. For the evolved m-rate regime, GA-M, the offline fitness is at best only comparable to, and frequently worse than, that of a random population (generation 0 of figure 2). The reason is that the population is converging around a point that lags the optimum thus limiting

a) Generation 300 offline fitness

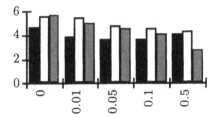

b) Generation 300 online fitness

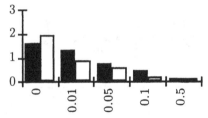

c) Time averaged equilibrium best fitness

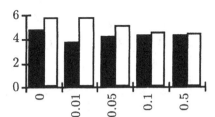

d) Time averaged equilibrium mean fitness

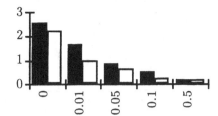

Figure 1. Fitness values against environmental rate, a, for GA-M (solid), SGA (empty), and Cobb's reported results (hashed) a) Mean offline fitnesses at 300 generations; b) mean online fitness; time averaged best (c) and mean (d) fitnesses taken over the last 126 generations.

the available variation and as a result, offline fitness. Note how the steady state time average values offer a clearer picture of the equilibrium behaviour.

Whilst the improvement in online fitness of GA-M over the fixed, 'optimal' rate may seem slight, just how significant the different approaches are in terms of the ability of the population *as a whole* to track the environment is revealed by the behaviour of the mean population *phenotype*, figure 3, which depicts a single run (the behaviour being typical); see also figure 1d. The SGA requires a high, fixed mutation rate to source enough variation for the efficient *offline* tracking of a rapidly moving environment; however, this high m-rate has a detrimental effect on the ability of the population as a whole to

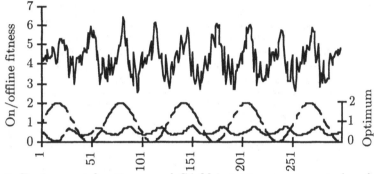

Figure 2. Environmental optimum and GA-M instantaneous mean and peak fitness against generation number for a = 0.1.

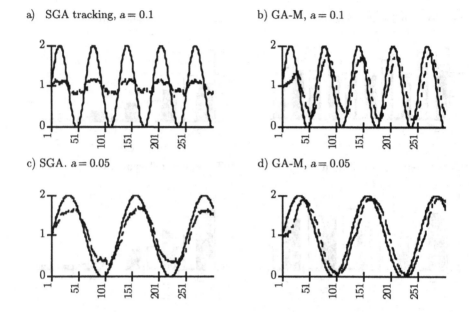

Figure 3. Tracking a moving environment over 300 generations- a) and b), a = 0.1; c) and d), a = 0.05. In all cases, the trajectory of the environmental value over time is very closely followed by the best of each generation. The solid line represents the environmental target, almost ideally tracked by the best of each current population. The dashed lines represent the mean phenotypic value of the population during the current generation - a) fixed m-rate of 0.5; b) evolved m-rate + 0.001 background mutation; c) fixed m-rate of 0.1; d) evolved m-rate + 0.001 background mutation.

track the environment, figure 3a. For the case of GA-M, whilst offline performances are down (though not noticeable on the trace shown), the population mean is able to track the environment far more efficiently, figure 3b. In a slower environment, where a lower rate of mutation is applied, mean tracking behaviour is much improved in both cases; and although the online and offline fitnesses are worse for GA-M compared to SGA, the final generation population convergence (measured as mean proportion of the majority bit at each locus over the population) is higher (figure 4) and mean tracking better for all values of a, with convergence increasing for decreasing mutation rate.

Looking at the instantaneous mean and peak fitnesses for GA-M for a particular environmental rate, (specifically, a = 0.1, figure 2), the scores oscillate with a period half that of the environmental signal (i.e. twice the frequency) with the mean performance lagging the peak performance. Comparison of the phase of these fitness traces with that of the environmental optimum $h(t)$, suggests that the population best fitness peaks just after the rate of change of $h(t)$ is at a minimum (i.e. when the absolute value of the gradient of $h(t)$, that is $\cos(at)$, is at a minimum). The population mean fitness, however, lags further still.

As far as the actual fitness values go, once the system has settled the time averaged means seem to be on a par with the scores recorded by the initial random population, although the degree to which the corresponding populations are converged is significantly different (figure 4).

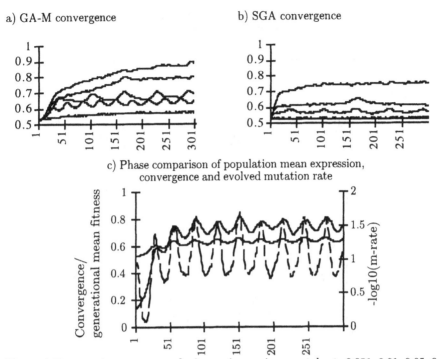

Figure 4. Decrease in convergence for increasing environmental rate 0.001, 0.01, 0.05, 0.1, 0.5: a) regime GA-M; b) regime SGA; c) phase comparison of convergence (middle trace), population generational mean fitness (lower trace) and (-log10) mutation rate (upper trace, right hand scale) for GA-M, a = 0.1.

Turning now to the degree to which the populations converge. Firstly, it appears that in general the degree to which a population is stably converged (i.e. the equilibrium convergence value) falls with increasing *environmental* rate. However, inspection of figure 4b suggests that in fact the degree of convergence is actually reflecting the mutation rate since for the two fastest moving environments, where the *same* fixed mutation rate is implemented, the convergence levels are similar. For the rapidly changing environment, the initial increase in convergence is largely a result of the convergence of the rate gene (convergence is measured over *all* genotypic bits).

It is interesting to compare the phase of generational mean fitness, convergence and mutation rate for a single environmental condition, figure 4c. In particular, the mean and convergence traces are in phase, whereas the mutation rate is 180 degrees out of phase (since the higher the value of -log(mutation rate), the lower the actual rate). With the mutation rate at a minimum, convergence and mean fitness are close to their maximal values.

Figure 5 shows the mean instantaneous evolved mutation rate over a and table 1 compares the time averaged final generation mutation rate to the values selected by Cobb. Clearly, her intuition was correct in that higher mutation rates are required to track the environment successfully: increasing a increases the evolved mutation rate, although only as much mutation as is necessary is supported. The far lower evolved rates explain the better mean population tracking behaviour of GA-M. Note how for intermediate a the raw evolved mutation rate (figure 5) oscillates at twice the environmental rate as did the instantaneous mean and peak fitnesses.

Table 1. Evolved mutation rates over a with 4 additional rates (bracketed); mean rates given are taken over the whole run and the last 150 generations (this latter representing an equilibrium rate). For GA-M, there is an additional background rate of 0.001/bit.

a	0.001	0.01	(0.025)	0.05	0.1	(0.25)	0.5	(1.0)	(5.0)
SGA	0.01	0.05	N/A	0.05	0.5	N/A	0.5	NA	N/A
GA-M:									
1-300	0.023	0.031	0.034	0.044	0.061	0.136	0.238	0.287	0.231
150-300	0.002	0.004	0.008	0.016	0.032	0.102	0.194	0.200	0.128

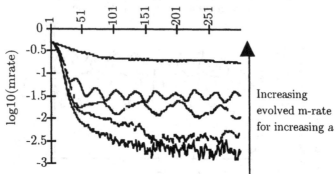

Figure 5. Mean instantaneous evolved mutation rates against generations for GA-M, a = 0.001, 0.01, 0.05, 0.1, 0.5. The arrow depicts increasing alpha and mutation rate.

Having evolved mutation rates, it makes sense to see what exactly is being optimised by utilising a fixed mutation rate of similar magnitude to the evolved rate for a particular a. Table 2 shows online and offline fitness values for fixed mutation rates in the vicinity of the evolved mutation rate for a = 0.1 (specifically, the time averaged evolved rate was 0.062/bit taken over generations 1-300, 0.032/bit over generations 150-300 (this latter value being the equilibrium rate)). Whilst offline fitness improves for increasing fixed m-rate, as an increasing amount of variation is available, the online performance peaks for m-rate 0.045, the mid-point of the two time averaged evolved values given in table 1.

Further consideration of traces shown in figure 3 reveals both an attenuation of, and a phase lag between the mean value expressed by the population and the environmental value. I have already noted how increasing the mutation rate affects the stability of the population and its ability as a whole to track the environment - for higher mutation rates, the mean phenotypic value of a population fails to track the environmental target.

Table 2. Online and offline fitness versus fixed mutation rate in the vicinity of the evolved rate for a = 0.1, selection rank minimum 0.5.

m-rate/bit	0.015	0.030	0.045	0.060	0.075
Online fitness	0.49	0.55	0.56	0.55	0.53
Offline fitness	2.51	3.11	3.54	3.81	4.06

Changes in gain (or attenuation - that is, the peak-to-peak value of the tracking signal compared to that of the optimum) and lag (the phase difference in generations between tracking and tracked signals) result from altering the genetic variance, as regulated by the mutation rate. Table 3 shows how phase lag is affected by mutation rate. Mean values are mean delays in generations between related target and population mean values passing through 1.0, discounting the first proper cross, and taken over 10 independent runs. Specifically, lag is decreased by a slight increase in the mutation rate (0,01 on a rate of 0.045), although only by a generation or so. For larger changes in the mutation rate, tracking properties of the population are altered and useful direct comparison becomes difficult. For larger changes in the mutation rate, tracking properties of the population are altered and useful direct comparison becomes difficult. The amplitude of the mean expressed phenotype is similarly a decreasing function for increasing mutation rate, although the effect on mean tracking ability is the converse to that of the lag. That is, as mutation rate increases, whilst the lag between mean phenotype and the environmental target is reduced, the attenuation of the mean increases (i.e. there is a *worse* fit between the time delayed environmental signal and the mean value).

The gain of the population mean against environmental frequency (not shown) has the form of a low pass filter. This is in accord with theoretical results [3], and is discussed in more detail in [4].

As with mutation rates, so with selection pressure, parameterised by the rank minimum value: by increasing the selection pressure, the effectiveness of a population's mean phenotypic tracking improves (not shown) and phase lag decreases, as demonstrated in table 4.

Table 3. Regime SGA - selection rank minimum fixed at 0.5; a = 0.1; discard first cross after origin start; change mutation rate from 0.015 to 0.075 step size 0.015.

m-rate/bit	0.015	0.030	0.045	0.060	0.075
peak to peak	1.71	1.64	1.56	1.45	1.38
mean lag	7.70	6.39	5.74	5.15	4.80
lag sd	0.44	0.38	0.32	0.27	0.36

Table 4. Regime SGA - mutation rate fixed at 0.045; a = 0.1; discard first cross after origin start; relax selection, parameterised by rank minimum, from 0.1 to 0.9 step size 0.2.

rank min.	0.1	0.3	0.5	0.7	0.9
Online fitness	0.81	0.70	0.56	0.39	0.19
Offline fitness	3.39	3.48	3.54	3.71	4.09
mean lag	4.00	4.80	5.74	7.28	8.75
lag sd	0.27	0.28	0.32	0.21	0.99
peak to peak	1.75	1.68	1.56	1.29	0.59

3 Tracking Fluctuating Environments with Plasticity

In this section, I report on an individual adaptive strategy that employs a form of trial and error learning. A regime in which the inheritance of characters acquired by means of this strategy is also considered.

3.1 Modeling Plasticity

Plasticity, herein taken as a simple form of adaptive development or learning (i.e. within generation, local adaptive search), allows any particular to genotype to realise one of several possible phenotypes. Plasticity is modeled through a probabilistic mapping between each genotypic locus (bit position) and its phenotypic correlate over a fixed number of (learning) trials. In each learning trial, every 'phenotypic' locus is set to a value equivalent to the inherited allele with a probability (1-*rate*), its complement otherwise, where *rate* is an evolved learning rate (1-rate) carried independently by each individual. The evaluation value returned for each individual is then the lowest error score achieved over all that individual's trials. Note the 'innate' evaluation of the individual (i.e. the raw evaluation of the pre-developed individual) is not measured unless no bits happen to flipped in a particular trial. Whilst this scheme may seem to offer cost free plasticity, the probabilistic nature of the learning regime means there is an implicit cost between generations through the imperfect transmission of the parent's adaptive phenotype (discounting transmission errors arising from recombination). Cultural inheritance (through imagining the model at a solely phenotypic level, equating the inherited state to an initial phenotypic state rather than as a genetic message), or the inheritance of characters acquired through plasticity in a developmental model, is achieved by transmitting the genotype of the developed individual that gave rise to the lowest evaluation. IAC is applied to the fittest learned phenotype and is achieved by copying the phenotypic bits directly onto that individual's genome.

3.2 To Compare SGA and Plastic GA With and Without the Inheritance of Acquired Characteristics (IAC).

In this experiment, I compare the performance of a plastic GA (using the scheme described above) with and without IAC (GA-P and GA-IAC respectively) with SGA from the previous experiment in the same sinusoidally fluctuating environment of equation (1). For each individual, 10 learning trials are allowed; the learning rate is an individually carried, evolved probability between 0.0 and 0.5[1] (the same rate for all bits, but an independent rate for each individual), represented as 5 contiguous binary coded bits. In both GA-P and GA-IAC regimes, there is a background mutation rate of 0.001 applied to *all* bits. Population sizes for these two learning regimes were set at 100, half the size of the population in the mutation only conditions.

Comparing generation 300, time averaged fitness data, figure 6, both plastic regimes offer best of generation individuals that perform at least as well as for the SGA in all environments, and considerably better in the faster changing ones. In addition, mean online performance levels are higher in all environments for the plastic cases, GA-IAC outperforming GA-P. Note too that the mean online fitness for GA-P is similar across all but the slowest moving environment, whereas the mean fitness for GA-IAC deteriorates.

How does this translate in terms of actual tracking ability (figure 7)? For the cases of plasticity, there are three useful measures that may be taken:- the best and mean phenotypic expression of the population, and the population

---

[1] Setting the range between 0.0 and 1.0 allows each individual to choose its polarity. For the sake of simple measurements, the rate was thus restricted to half the unit range.

a) Generation 300 offline fitness

b) Generation 300 online fitness

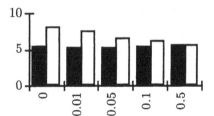

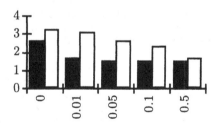

Figure 6. Fitness values against environmental rate, a, for GA-P (solid), GA-IAC (empty); a) Mean offline fitnesses at 300 generations; b) mean online fitness.

a) GA-P tracking, a = 0.1

b) GA-IAC tracking, a = 0.1

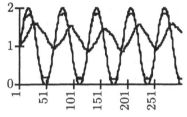

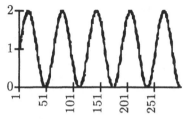

c) GA-P tracking, a = 0.05

d) GA-IAC tracking, a = 0.05

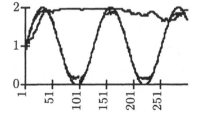

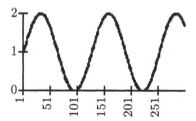

Figure 7. Tracking a moving environment during a single, typical run over 300 generations- a) and b), $a = 0.1$; c) and d), $a = 0.05$. In all cases, the trajectory of the environmental value over time is very closely followed by the best of each generation. The solid line represents the environmental target, almost ideally tracked by the best of each current population. The dashed lines represent the mean phenotypic value of the population during the current generation - a, c) GA-P; b, d) GA-IAC.

mean direct evaluation of the genotype without plasticity. As for the mutation only cases (SGA, GA-M), the best individual of each generation tracks the target well, with the population capable of IAC closely following the environment as a whole both at the expressed and innate level. Specifically, closer inspection reveals that for GA-IAC in the slowly changing environments there is a lag of a single generation between environment and mean expressed value and two generations between environment and mean innate value; as the environmental rate increases, the population expressed mean tracks without any lag and the innate mean lag is reduced to a single generation. Similarly for GA-P, the population mean expressed phenotype lags the environment by a single generation in slow environments, not at all in faster ones, but the directly evaluated value lags considerably further (in the order of 10 generations for a = 0.1), greater in fact than for the case of SGA. This is in accord with a quantitative genetics analysis of Anderson [7] in which plasticity is modeled by a relaxation of the selection strength in both static and dynamic environments. In any particular run, the low rate at which *genetic* variation is introduced into the gene pool means the population may converge to a point from which learning alone is sufficient to track the moving target. In other cases, however, some degree of gene pool tracking is supported, although as has been mentioned, the lag is considerable. This problem may be addressed by increasing the background m-rate.

As with the evolved mutation rate, regime GA-M above, higher environmental velocities support increased amounts of introduced variation, whether through an increased learning or mutation rate. Looking at the evolution of the learning rates, for regime GA-P, figure 8a, the rate assumes a relatively stable value for high and low environmental velocities, but for intermediate rates of change the rate itself varies as the population as a whole tries to track the environment. In a numerical simulation involving environments that move at a constant velocity, [7] reports a similar increase in the amount of learning supported for increasing environmental rates. Where IAC is allowed, figure 8b, stable, lower learning rates are quickly adopted across the range of environments, although for intermediate a there is an initial period of increasingly damped oscillation. For a = 0.1 at least, the rate evolved under GA-IAC is little affected by the strength of selection (not shown).

a) GA-P l-rate

b) GA-IAC l-rate

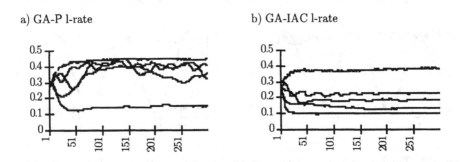

Figure 8. The evolution of learning rates over the range of environmental velocities for a) GA-P; b) GA-IAC. Generally, the higher the value of a, the greater the learning rate.

Since the developmental map is essentially simple (phenotypic states have a direct genotypic coding), GA-IAC may be thought of as an example of directed mutation as opposed to the weaker adaptive mutation of regime GA-M. That is, only currently *useful* 'mutations' are allowed to occur. It thus makes sense to compare the evolved rates for these two regimes: in particular, GA-IAC supports a range of switching rates of comparable high order across all environments, whereas the evolved mutation rate of regime GA-M spans two orders of magnitude and only compares with the rates of GA-M in the fastest moving environment.

Considering the degree of convergence for the plastic populations over the range of environmental rate (figure 9), GA-P populations maintain a higher degree of convergence than do GA-IAC populations, as a result of the low rate

a) GA-P convergence

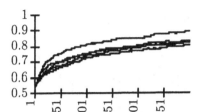

b) GA-IAC convergence

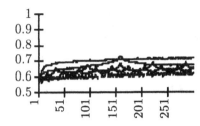

Figure 9. Convergence for a) GA-P; b) GA-IAC.

at which genetic variation is introduced. For intermediate to large a, the IAC population displays a degree of convergence on a par with that of the evolved mutation rate case, regime GA-M. For low values of a, GA-P converges to a similar degree as GA-M, although at higher fluctuation rates the plastic algorithm maintains a consistently high level of convergence with a high degree of plasticity supplying the phenotypic variation necessary for effective tracking and low mutation restricting the amount of genetic variation. It falls to further work to make a direct comparison of the convergence of an SGA regime and a GA-P regime with similar fixed mutation rates across all environmental rates.

4 Fluctuating Environments with Periods of Stasis and Injected Noise

In Cobb's original paper was a comparison of her mutation based algorithms in mixed stationary/non-stationary environments. Using the set up of the previous experiment, figure 10 depicts the tracking behaviour for each condition. For the non-plastic, fixed mutation rate case (SGA), figure 10b, the high mutation rate prevents the population from phenotypically converging to the optimal value when the environment becomes stationary. GA-P also has problems in terms of mean population performance at the start of the stationary period although it does improve. One of the reasons this takes so long is the high rate of learning (figure 11a) combined with the initially low level of convergence (figure 11b).

As expected, when the fluctuating environment becomes stationary, evolved rates fall (figure 11a), although again GA-P supports a higher rate than GA-IAC. For both GA-M and GA-IAC, convergence is rapid, mediated by convergent pressures (selection, recombination, little mutation) in the former case, IAC in the latter, and for which (genetic) variation may still be introduced through the not insignificant learning rate. The SGA population only converge slightly the stationary period because of its high, foxed mutation rate. While the environment is fluctuating, novel variation is introduced at a constant rate, and because the direction in which selection is acting keeps changing (the environment and hence the optimal genotype is non-stationary) the population is unable to converge on any particular point. When the fluctuations cease, selection is better able to locate useless variation and drive it from the population. When the environment begins to move again, GA-IAC and SGA begin to diverge before GA-M. For GA-P, the population converges steadily throughout, and pays little heed to the state of

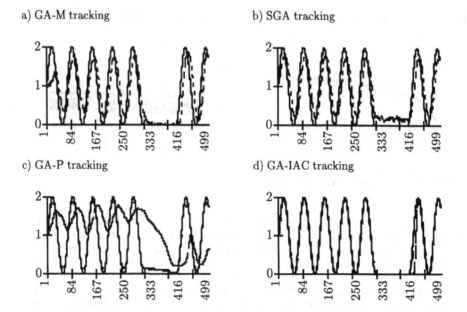

Figure 10. Environmental tracking in combined stationary/fluctuating environments: a) GA-M; b) SGA; c) GA-P; d) GA-IAC.

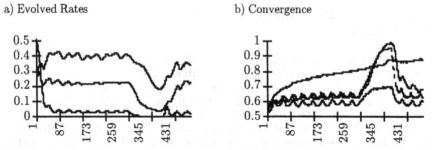

Figure 11. Evolved rates and convergence in combined stationary/fluctuating environments; a) evolved rates (GA-P highest, then GA-IAC, then GA-M); b) convergence (GA-P consistently increasing; during stationary period, GA-M highest, then GA-IAC, then SGA).

the environment, as the rate at which variation is introduced into the population remains constant and selection is weakened through plasticity. Continuing the run for GA-C does not increase convergence much more than its generation 300 level.

5 Fluctuating Environments with Injected Noise

Figure 12 demonstrates how each method copes with perturbations to the otherwise sinusoidally varying environmental target. The two mutation based algorithms maintain enough variation to track the target closely, but the immortalisation of the one time adapted states is such a slow process that the

a) GA-M tracking

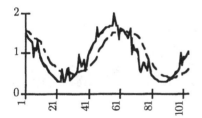

b) SGA tracking

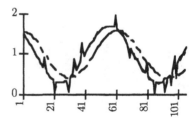

c) GA-P tracking

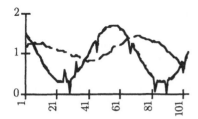

d) GA-IAC tracking

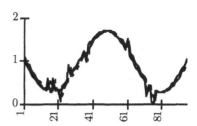

Figure 12. Environmental Tracking for Sinusoidally Varying Environment with Infrequently Injected Noise, $a = 0.1$, single run traces; a) GA-M; b) SGA; c) GA-P; d) GA-IAC.

mean phenotypic expression of the population maintains its smooth rate of change, as does the particular GA-P population shown. This is a good example of the low pass filtering effect of evolution discussed in [4]. GA-IAC so rapidly assimilates information into the germ line that both mean acquired and innate traces closely track the environment, including perturbations, albeit with a slight phase lag.

6 Coevolving Rates

With the addition of a second, evolvable rate parameter to the genome of each individual, it is possible to set the rate at which IAC occurs, in a regime denoted GA-C: the rate I gives the probability that each *acquired* bit will be transmitted, otherwise that individual's *inherited* state is transmitted. This experiment has a theoretical population genetics counterpart in a simple analysis by Boyd [8].

A second pairing of evolved rates that suggests itself is that of learning and mutation rates, GA-PM. For GA-P, it was found that the low background mutation rate occasionally resulted in the population failing to track the environment genetically and converging to a point from which plasticity had to provide all the necessary variation. By evolving the mutation rate, the population will be able to adapt the amount of variation introduced by that mechanism.

The results of running GA-PM and GA-C ten times over 300 generations are shown in figure 13; comparing this with figure 6, GA-C is seen to perform as well as GA-IAC in offline fitness terms, although underperforming slightly

a) Generation 300 offline fitness

b) Generation 300 online fitness

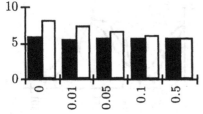 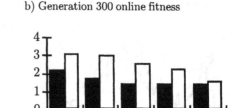

Figure 13. Fitness values against environmental rate, a, for GA-PM (solid), GA-C (empty); a) Mean offline fitnesses at 300 generations; b) mean online fitness.

with regard to online performance. GA-PM outperforms GA-P for offline fitness, with again a slight underperformance by the online fitness measure. Again, the plasticity with simple inheritance strategy offers similar mean fitness across all but the slowest environments, whereas mean performance falls off for increasing a where IAC is allowed.

Turning to a comparison of the evolved rates, figure 14; for GA-C, IAC is universally supported in all but the fastest moving environments, with an increase in learning rate for increasing alpha. For high alpha, the culture rate begins to drop and the learning rate increases greatly - for faster moving environments, learning predominates over culture. For GA-PM, in the slower environments, the mutation rate is low and the learning rate copes. As the environmental period decreases, the learning rate saturates at its maximal value and the mutation rate climbs to a high level. This is not necessarily an

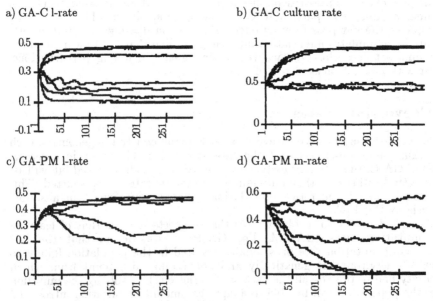

Figure 14. Evolved rates for GA-PM and GA-C (a rates as above, plus 1.0 and 5.0): a) GA-C l-rate (increasing for increasing a); b) culture rate for GA-C (decreasing for increasing a); c) GA-PM l-rate (increasing for increasing a); d) GA-PM m-rate (increasing for increasing a).

adaptive measure. If a high mutation rate is not detrimental, it will climb. The selection pressure, weakened by plasticity, is not great enough to restrict the genetic variation through suppression of the mutation rate.

Finally, in the noise injected environment (not shown) GA-C performs similarly to GA-IAC, as one might expect, but for GA-PM the mutation rates are so high that no direct tracking occurs.

7 Summary

This paper presents intermediate results from an ongoing study of the artificial evolution of populations of adaptive individuals in sinusoidally fluctuating environments [5]. Population members were allowed to evolve individual mutation or learning rates. Generally, as the environmental period decreased, the evolved rate increased. An evolvable m-rate allowed the population as a whole to track the optimal sequence rather more closely than for the fixed m-rate case.

For a mutation only strategy, the population mean expression represents a lagged and attenuated version of the optimum. It has been suggested elsewhere that this corresponds to the behaviour of a lowpass filter [4]. Removal of sporadically injected noise would appear to support this view. Where plasticity with simple inheritance is introduced, the expressed population mean tracks the optimum more closely, although the underlying genes lag ever more. With IAC, lag is reduced to a single generation and the underlying gene pool is itself able to track the optimum very closely.

One problem identified during the experiments reported herein was the suitability of various 'traditional' measures of fitness for sensibly reporting the behaviour of populations evolving in fluctuating environments. Where only steady state, equilibrium behaviour is of interest, time averaged measures that take into account the early generations may skew the result. Where knowledge of the settling time behaviour is also required, then traditional on/offline measures may be suitable.

Comparing single strategies (SGA, GA-M, GA-P and GA-IAC) over environmental rate, GA-IAC appears to offer the best response, at least in terms of the fitness measures reported here. GA-P offers the next best strategy followed by the mutation rate regimes with a suitably set mutation rate. However, GA-P does appear to guarantee mean performance across a, whereas for all other strategies, mean fitness deteriorates with increasing environmental rate.

Additionally, IAC itself may be costly both in terms of local search, and also in terms of 'reverse engineering an acquired phenotype in order to generate the genotype that represents it. Theoretical results from [8] suggest that in slow moving environments, mutation only strategies fare better than strategies involving costly plasticity. In environments with an intermediate period (10's of generations) cultural inheritance/IAC offers an optimal trade-off between individual learning and inherited traits. In rapidly fluctuating environments, where there is little correlation between adapted individuals of one generation and the next, an individual learning strategy is most appropriate.

By co-evolving plasticity rate and the degree of cultural transmission, regime GA-C, the trade-off identified in the literature between cultural inheritance and individual learning for varying temporal grain of the environment has been demonstrated. Specifically, as the environmental rate, a, increases, the learning rate increases but the inheritance rate falls.

The co-evolution of learning and mutation rates demonstrated by regime GA-PM is also of interest, since it suggests that mutation rates tend towards the highest value they can get away with, as opposed to the lowest.

Acknowledgments

This work was carried out under a PhD grant from the Open University. Thanks to Jon Rowe, an anonymous reviewer and Russell Anderson, for their comments on an earlier draft of this paper.

References

1. Cobb, HG. (1990) An Investigation into the Use of Hypermutation as an Adaptive Operator in Genetic Algorithms Having Continuous, Time-Dependent Nonstationary Environments. Navy Center for Applied Research in AI. December 11, 1990.
2. Dasgupta, D. (1993(revd 1994)) Optimisation in Time Varying Environments Using Structured Genetic Algorithms. Technical Report IKBS-17-93, Dept of Comp Sci, University of Strathclyde. December 1993 (updated 6/4/94).
3. Lande, R, & Shannon, S. (1996) "The Role of Genetic Variation in Adaptation and Population Persistence in a Changing Environment." *Evolution* 50(1):434-43.
4. Hirst, AJ. (To appear) Evolutionary Signal Processing: A Preliminary Study. Proc. ECAL97, MIT Press.
5. Hirst, AJ. (in prep) The Interaction of Evolution, Plasticity and Inheritance in Genetic Algorithms. PhD Thesis Dept. of Psychology, Open University.
6. Turney, P. (1996) How to Shift Bias: Lessons from the Baldwin Effect. Evolutionary Computation 4 (3) pp. 271-295
7. Anderson, RW. (1995) "Learning and Evolution: A Quantitative Genetics Approach." *Journal of Theoretical Biology* 175: 89-101.
8. Boyd, R, & Richerson, PJ. (1988) "An Evolutionary Model of Social Learning: the Effects of Spatial and Temporal Variation." *Social Learning: Psychological and Biological Perspectives.* Ed. TR Zentall & BG Galef Jr. Lawrence Erlbaum Associates. pp. 29-48.
9. Hirst, AJ. (1997). On the Structure and Transformation of Landscapes. This volume.

On the Structure and Transformation of Landscapes

Tony Hirst

Dept. of Psychology, Open University, Milton Keynes MK7 6AA UK

Abstract. Whilst the metaphor of 'fitness landscapes' is widely applied in the evolutionary algorithm (EA) community, there are several assumptions requiring its application that are often ignored, such as the underlying structure of the search space and the ontological status of the values depicted by the landscape. By differentiating between valuation and evaluation surfaces, and surfaces of selective value, it is possible to show how each may be transformed by learning (and learning costs) and the inheritance of traits acquired therefrom. In doing so, two styles of learning are identified, *rank respectful* and *rank transforming*, and these are shown to behave differently under proportional and rank based selection schemes.

1 The Notion of Landscapes

Virtually everyone in the evolutionary algorithm community is familiar with the notion of a *fitness landscape*, conceptualised as a surface of individual fitnesses plotted over binary genotypes one Hamming unit apart. Originally proposed by Sewall Wright [1] there is historically a certain degree of confusion as to the actual nature of the space over which the landscape lies [2]. The height of the surface at a particular point represented either the fitness of the individual specified by that point, or the mean fitness of the population. The former representation of the search space presented a gross simplification of the actual, high dimensional space, typically by assuming that neighbouring individuals were separated by a single mutation. The latter view required a search space describing all possible gene frequencies, in which each point gave the gene frequencies over a population. As Wright points out, the device is a "pictorial representation" invented to illustrate his Shifting Balance theory of evolution, that is "useless for mathematical purposes". However, it does present an intuitive view of the relative fitness values between genetic neighbours and is frequently used today by those in the evolutionary community as a 'fitness' value over a single representational axis.

Landscapes in Evolutionary Computation. In the field of evolutionary computation (EC), there has been considerable confusion between the role of the objective (evaluation) function to be optimised and the fitness function that represents the evaluation function transformed through selection and sampling. The two are often assumed to be one and the same, as for example in the Goldberg's widely used textbook [3], but as I shall demonstrate, this is not the case. In biology too, there are certain 'philosophical problems' with the notion of fitness (e.g. [4][5]), and it may be that a clearer understanding of fitness in evolutionary computation terms may shed some light on the biological position.

In the simple model of the 'within generation' components of a simple genetic algorithm, figure 1, an individual genotype, G, receives an evaluation, E, directly (as in the case of the MaxOnes (bitcounting) evaluation function);

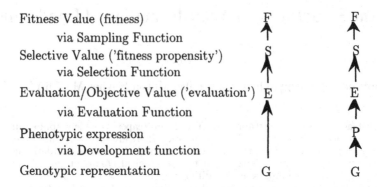

Figure 1. The *within generation* functional components of a canonical EA.

or according to its phenotypic expression, P, derived through some development function. An example of the latter case is the MaxInt evaluation function, where the aim is to maximise the integer value of an individual phenotype that is suitably coded for by the genotypic bitstring.

In simple optimisation models, all the determinants of selective value are known (i.e. the evaluation and selection functions and the optimisation criteria). The *selective value*, S, itself may be thought of as a 'fitness propensity', or 'prescriptive fitness', since it defines the probabilistic distribution of selected parents. The 'actual' or 'real fitness' (fitness, F) of an individual is then a simple function of the *actual* number of times it is sampled for breeding. I shall use the term *performance surface* to refer to the general class of performance values that may be usefully thought of in terms of landscapes.

In a *proportionate based* selection scheme (PBS), the selective value of an individual *is* equivalent to its evaluation, normalised by the population mean evaluation:

$$S = \frac{E}{\overline{E}} \tag{1}$$

This does not generally hold for *rank based* selection (RBS), where the selective values of individuals are defined in relation to the other individuals making up the current population through a ranking operator.

Selective value and fitness are equivalent if an ideal sampling function is employed, although typically two sorts of noise will be introduced during sampling: one due to finite population size (i.e. required proportions are not supported by the finite population size), and the other due to sampling error. Under the assumptions of proportional selection and ideal sampling, it is possible to save the use of the word *fitness* as referring to individual *evaluations*, since under such circumstances fitness is directly proportional to evaluation. However, this is not strictly true for RBS, where rather than reporting *fitness* measures, best and mean *evaluations* are typically presented.

Generally, the fitness function represents the evaluation function combined with a selection function and a sampling algorithm. This approach also suggests the possibility of an additional performance measure, specifically the mean evaluation of the *selected* population, in contrast to the mean evaluation of the population as a whole. This may be of interest in that it allows one to measure the gross correlation between the mean evaluation of selected parents and the mean evaluation of the population they give rise to through recombination. Such measures frequently offer insights into the suitability of given operator sets to particular evaluation functions (see for example, [6]).

In this paper, I shall further develop the distinction between stationary evaluation and selective surfaces, showing how they may be transformed through the choice of selection operator or the introduction of some 'learning' operator that may or may not act with some cost to an individual's evaluation. In addition, the transformation of population structure through the inheritance of acquired characteristics (IAC) will be discussed.

2 Surveying the Landscape

Whilst the idea of landscapes is pervasive in the evolutionary computation literature, there are three assumptions critical to the sensible application of the metaphor that are often ignored.

First, as I suggested above, the landscape idea is typically used to help visualise the motion of a population as it evolves over a stationary *evaluation* landscape, where individual evaluations are taken to be independent of each other. However, one should also bear in mind the *selective surface* which represents the evaluation landscape over a population transformed by the particular selection function being used.

Second is the identity of the search space - Provine [2] points out that Wright is not consistent in his view of what the search space underlying the landscape corresponds to. For example, at some times it is taken to be the space of all individual genotypes, at others to be the gene frequencies in the population.

The third assumption relates to the underlying structure of the hyperplane over which the surface is plotted, that is, what connects neighbouring points? Wright himself argued that the multidimensionality of the population space "should be explicitly assumed to underlie the two-dimensional one of the diagram and that its origin should be at whatever peak was under consideration" ([1] p120). Elsewhere, I have offered a pragmatic simplification to a visualisation of the whole search space, which concentrates on the search space neighbourhood (SSN) of a population induced by the particular operators being applied [8]. The view is further complicated by considering developmental (genotype-phenotype) mappings and crossover. Evaluation is usually applied to phenotypes, but the evolutionary operators of recombination act at the genotypic level. One must be clear to distinguish, then, between the evaluation surface plotted over genotypic and phenotypic representation spaces (Sect. 2.4).

Having clarified the landscape type and structure, it is possible to use plasticity (e.g. learning) and selection operators to transform and map between the various types of landscape.

2.1 'Fitness Landscapes': Surfaces of Evaluation or Selective Value?

The selection strategies used in evolutionary algorithms (EA's) fall into one of two classes - proportionate based selection or ordinal (rank based) selection [8]. Proportionate schemes select individuals in proportion to their evaluation, often normalised according to the mean evaluation of the population. Rank based schemes order the population in terms of evaluation, and select individuals according to their rank within the population. Changes in evaluation (such as those arising from learning) affect these two types of selection in different ways. According to [8]:

> "A selection scheme is said to be scale invariant if multiplying the individuals' fitness by a constant does not change the selection pressure. A selection scheme is said to be translation invariant if adding a constant to every individuals' fitness does not change the selection pressure. Proportionate selection methods are normally scale invariant, but translation variant. Ordinal-based selection schemes are translation and scale invariant."

By considering the *selective values* afforded to individuals under selection, this distinction is made apparent in terms of the range of possible selective values. Under proportional selection, the range of selective values depends on the normalised evaluation of each individual and reflects those individual evaluations. For RBS, for a fixed population size, the set of actual selective values awarded during any generation is the same, whatever the population[1]. It is not surprising, therefore, that the idea of a *fixed* 'fitness landscape' has come to represent *objective* values or *evaluations*, not least since this surface may be constructed over the whole of the genotypic space. Note, however, that this affords little information as to the selective value, or relative evaluation, of a given individual in an arbitrary population under RBS. I suggest that in general a discussion of populations evolving through depicted or imagined performance landscapes under selective conditions mentioned only in passing may be improved by also bearing in mind the selective landscape over the current population or temporally local, expected populations.

Example 1.1. Imagine a set up utilising truncation selection with elitism and a threshold of $x\%$ (that is, the 100 - $x\%$ worst performing individuals in a population with N members are discarded (i.e. their selective value is 0)), selecting individuals from the remaining $x\%$ with equal probability, $1/(x\% * N)$, with the proviso that the 'fittest' (elite) individual is always selected and placed into the selection pool. The MaxInt evaluation function over the current population may now be transformed into a function that defines a selective surface. The value of the surface over an individual represents the likelihood of it being selected for the next generation. This transformation makes concrete the distinction between evaluation and selective values. See, for example, figure 2, which considers the evaluations in a population of 10 binary coded individuals, $-1, 3, 5, 6, 6, 7, 8, 9, 11, 13$", for the 4 bit MaxInt problem, and the selective surface over the population under proportional selection (figure 2a) and elitist truncation selection with threshold 50% (figure 2b).

2.2 Defining the Search Space

In what follows, I shall distinguish a range of search spaces over which one may visualise various performance surfaces: the *genic frequency* space (in which each dimension corresponds to the proportion in the population of

---

[1] It is assumed throughout that where two individuals receive the same evaluation, there is some mechanism for distinguishing between them so that the population may be completely rank ordered. This is likely to introduce extra noise into the system, as for example where the similar individuals are randomly ranked with respect to each other.

binary alleles set to 1 at each locus; the *individual genotye* space, in which each node corresponds to a unique genotypically defined individual; the *genotypic population* space, where each point represents a particular, unique population; and the *individual phenotype* space, where each point represents a distinct, phenotypically defined individual. (A summary of these spaces for the non-developmental case, (where strictly the development function is the identity) is given in Table 1). It should become obvious that the atomic points characterising each space constrain the choice of performance surface that one may display over that space. Note that an additional time dimension is required if the temporal evolution of the population is to be displayed.

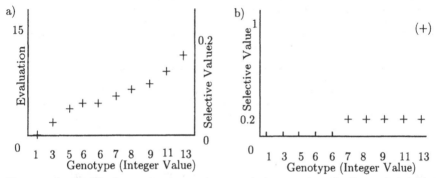

Figure 2. a) 4 bit MaxInt evaluation against population members for Example 1.1; b) the distribution of selective values for the same population under elitist truncation selection, with cutoff 50%.

Throughout the rest of this section, I assume a fixed population size, containing N individuals, each with a binary genotype of length L bits, and independent evaluation of individuals.

Genic Frequencies. This sort of space is typically used for plotting population mean evaluations; the evaluation of individuals and mean evaluation of particular populations is only recoverable if genic (individual locus) evaluations are additive. Mean selective value under RBS has no sensible application over this space.

Individual Genotypes. This space is suitable for identifying the independent evaluation of every possible genotypic individual. As each individual is represented uniquely, non-additivity of genic evaluations is supported. Selective values arising from RBS are not representable over this space.

Given a performance surface over this space, it is possible to generate the corresponding mean for a given set of gene frequencies or a given population. The reverse mapping (from means to individual evaluations) is not possible unless there is additivity of individual genic evaluations.

It is possible to represent a population by augmenting the space in the following way: associate with each genotype an integer quantity with the range 0...N. Setting this represents the number of individuals in a population bearing that genotype. There is then a straightforward mapping from the genotypic population space onto this augmented individual genotype space.

Genotypic Populations. This search space will happily underlie a surface representing population mean evaluations. If individual evaluations are independent, the evaluation of an individual is given by the mean evaluation of a population (size N) containing N similar individuals[2]. As for evaluations over genic frequencies, the mean evaluation of a population may be derived from individual genotypic evaluations, although again the reverse mapping only holds for additive genic evaluations. Note also that for an infinite population, the genotypic population space maps surjectively onto the space of genic frequencies.

It is possible to associate with each population point a subspace of the individual genotypic space that corresponds to the individual genotypes in the given population. The selective values of each individual under RBS may now be represented over this smaller embedded space for each population.

Table 1. Example search spaces that may underlie a 'fitness landscape'. in an algorithm *without* development.

Search Space Nodes	Volume of search space	Surfaces Evaluation	RBS	May be used to generate
Genic frequencies	L real dimensions	Population mean (1)		(2)(3)(4) if additive
Genotypic (individual genotypes)	2^L	Individual (2)		(1)(3)(4)(5)
Genotypic Populations [9]	$\left(\begin{array}{c} N + 2^L - 1 \\ N \end{array} \right)$	Population mean (3)	N/A	(2)(3)(4)
Genotypic Populations including explicit individuals	$2^L \cdot \left(\begin{array}{c} N + 2^L - 1 \\ N \end{array} \right)$	Population mean (4); Individual (5)	N/A	(2)(3)

Individual Phenotypes. It is impossible to find a *general* expression for the volume of the phenotypic space, as this depends on the developmental mapping from the genotypic space. However, where the mapping is known it *is* possible to calculate the size of these space. For example, to visualise a surface over individuals comprising a sequence of 5 Gray coded integers, each in the range 0 7, one requires a search space containing 8^5 points to characterise each individual. For an M point permutation problem, (e.g. the Traveling Salesman Problem), there are $M!$ possible permutations (phenotypic individuals), which is far smaller than the combinatorial search space fixed by the $0...(M-1)$ range of each of the M genes, which contains M^M points.

2.3 The Underlying Structure of Landscapes

There has been much interest of late on the structures that form the basis of an evolutionary search space. A view that is increasingly important is of a

---

[2] If the population structure affects individual evaluations, then the mapping from population to individual spaces may still hold, although it is likely that the reverse mapping will not. Note that in such a case, as for non-additive genic evaluations and the genic space, the mean population evaluation is not necessarily a useful measure.

search space represented by a graph whose nodes represent individuals, or pairs of individuals, and whose edges represent operator defined connections (e.g. [10][11][12]). Typically, the search space a landscape is visualised over is the Hamming graph, in which binary individuals are a single Hamming unit away from each of their neighbours. This space is suitable for representing a space traversed by an exactly odd bit flip mutation operator, but where crossover between individuals is supported, the definition of neighbouring points becomes rather less clear.

Even in the asexual case, the Hamming one graph may not appropriately represent points connected by mutation. For example, where an exactly even bit flip mutation operator is used (flips exactly 2, 4, etc. bits per individual), the binary space is partitioned into two halves, one containing strings with an even number of 1's, the other containing strings with an odd number of ones. Using such an operator, if the initial population contains individuals bearing only an even (or only an odd) number of ones, *in principle* only half the search space may be visited.

2.4 Transforming Landscapes Through Development: The Genotype-Phenotype Map

The *developmental map* is responsible for relating genotypes and the phenotypes that usually serve as the basis for their evaluation [13]. Typically, the map used in EA's is one-one (e.g. for evaluation functions defined over the integers, the developmental map will be binary or Gray coding), although this isn't *necessarily* so.

The opportunity to clarify further the notion of evaluation landscapes now presents itself, since we must decide whether the landscape refers to the surface over the phenotypic or genotypic search spaces. Following [14], I suggest that the objective values associated directly with the genotype of an individual are termed *valuations*, whereas the *evaluation* is the evaluation of the particular phenotype. So for example, in the case of MaxInt, where the evaluation function is defined over phenotypic integers, the *valuation* landscape may be transformed by different developmental maps (e.g. for a given genotype, its valuation will vary depending on whether the map is binary or Gray coding), but the evaluation landscape remains the same. Where a non-identity map is employed, the surface visualised over the spaces identified in Table 1 is the *valuation* rather than the *evaluation* landscape.

3 Transforming Landscapes Through Learning

The idea behind introducing a learning, or 'within generation, local adaptive search', operator is to improve the evaluation of each individual. The effect is that the evaluation function and the surface of selective values over the fixed population structure may be transformed. In the limiting case of within lifetime search to a local optimum, the evaluation of each individual is collapsed onto the evaluation of the local optimum in whose basin of attraction (induced by the learning operator) the individual lies. This approach corresponds to the memetic algorithms described theoretically by [15], depicted in figure 3 (based on [16]) and stands in contrast to the collapsing of a fixed learning neighbourhood, as described next.

In their widely reported paper on the interaction of learning and evolution, Hinton & Nowlan [17], hereafter H&N, introduced a simple learning scheme that incorporates several interesting properties. An individual

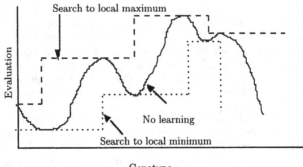

Figure 3. Transformation of the valuation landscape by learning and fault induction to local maxima/minima (adapted from a figure by [16]).

was represented by a string of 20 loci, each of which could take on the value **0**, 1 or **?** with initial probabilities 0.25, 0.25 and 0.5 respectively. The task was to match a target sequence of all 1's. For individuals containing one or more **?**'s, random 'learning' was allowed. A learning trial comprised of setting all the **?**'s in a particular individual to 1 or 0, each with probability 1/2, with a maximum of 1000 learning trials per individual. Learning stopped immediately if the target sequence was attained.

In this model, adaptive plasticity effectively replaces an individual by a virtual population of individuals, evaluation of the individual being based on the 'fittest' member of that virtual population. Furthermore, two parameters govern the extent of learning applied to an individual: the *learning distance* (that is, the scope of the learning neighbourhood given by the number of **?**'s in an individual) and the *number of learning trials*. These parameters are discussed in more detail below (Sect. 3.1).

More generally, two further things should be taken into account: a) what governs which population members may make use of this operator; and b) at what level of representation (genotypic or phenotypic) is the learning operator is defined? That is, does the operator define moves through the genotypic or phenotypic spaces? For example, in the binary coded TargInt function[3], a learning function may be defined at the genotypic level (e.g. single bit Steepest Ascent learning on the genotype) or the phenotypic level (e.g. add or subtract one to the integer value of the individual).

It should be remembered that for learning operators that may be defined in terms of local search at the genetic level, such as Steepest Ascent (SA) learning, the transformational effect may be different where different developmental functions are defined. This point will be illustrated in Sect. 3.3 where the effect of SA learning on the MaxOnes and MaxInt evaluation functions will be compared.

Where the learning function is deterministic (i.e. a given individual always learns the same thing), it may be possible to respecify the original evaluation function in an equivalent algorithm with a fixed evaluation function and no learning; for example, [16] have shown how it is possible to derive an equivalence evaluation function for a simple algorithm *without* learning that

---

[3] TargInt is the general case version of MaxInt, where the idea is to find the prespecified target integer (MaxInt defines the target as the *largest* integer in a given range).

behaves similarly (i.e. identically) to a partially deceptive evaluation function under Steepest Ascent learning.

Finally, by considering the population as a whole, and with learning available to every individual, two 'styles' of learning may be identified with respect to the relative evaluations of individuals before and after learning: *rank respectful* learning and *rank transformational* learning. The transformational effect of each style of learning then depends on the particular type of selection (rank based or proportional) being used (Sect. 3.2).

3.1 Learning Parameters

In this section, I shall consider in more detail the parameters relating to 'learning distance' and the number of learning trials.

Learning Distance: Defining the Learning Neighbourhood. The particular class of learning operators that I am considering are, like the mutation operator, applied independently to particular individuals. As such, each learning operator defines a 'learning neighbourhood' local to each individual. The size of the learning neighbourhood may be thought of in terms of how 'far away' from the inherited phenotype an individual may learn. So for example, in the model presented by H&N, the number of queries in any particular genotype specifies a learning neighbourhood for each individual, such that an individual bearing q ?'s has a learning neighbourhood that comprises 2^q members. If a ? had a 'default setting' in the event of the number of learning trials equaling zero, then q would represent the maximum learning distance (in Hamming units) away from the inherited, default phenotype that an individual could explore. Figure 4 demonstrates how the originally specified needle-in-a-haystack evaluation function is then transformed by adaptive plasticity into an *expected evaluation* function that varies smoothly over q.

Learning Intensity. How extensively an individual may explore its learning neighbourhood depends upon the number of learning trials.

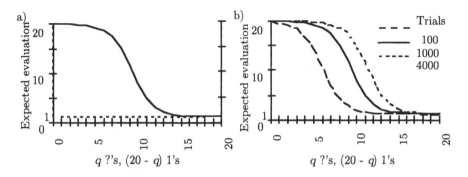

Figure 4. Transformation of an impulse evaluation function by learning for the evaluation function of H&N. a) the original experiment, in which the maximum number of trials is 1000, based on an analysis by [18]; b) for different limits on the maximum number of trials.

Note that even if the learning neighbourhood is searched exhaustively, the best evaluation returned may not be locally optimal (i.e. with respect to a repeated application of the particular learning operator). Rather, the best evaluation returned through an exhaustive search of the learning neighbourhood of an individual is the *neighbourhood optimum* for the individual undergoing learning; where the number of trials is such that the neighbourhood is not sampled exhaustively, the best evaluation returned is a *sampled neighbourhood optimum* for the given individual. If a random (learning) sampling strategy is followed, increasing the number of trials may well improve the performance in the H&N example, figure 4b.

3.2 Learning and Its Effect on Rank Ordering

The function 'transformations' that are of particular interest here are those that alter the (relative) selective values of individuals in a population. I distinguish two forms of transformation: *e-transformation*, a transformation of the evaluation function in terms of re-ordering the rank of individuals; and *s-transformation*, transformation of the selective values of an individual which maintains the rank order within the population. Table 2 summarises the susceptibility of PBS and RBS to these forms of transformation.

By defining the direct evaluation, e_g, of an individual, g, and the *benefit* from learning, b_g, received following learning by that individual, the evaluation, E_g, of g as passed to the selection function is given by:

$$E_g = e_g + b_g \tag{2}$$

The conditions for e- and s-transformation may now be given as follows:

e-transformation. If, for any individual, g, in the current population, the rank of an individual before learning is different to its rank after learning:

$$Rank\left(e_g\right) \neq Rank\left(E_g\right) \tag{3}$$

where $Rank(h_g)$ is the rank of an individual, g, with evaluation h_g, in the current population.

s-transformation. If, taken over all individuals, g, in the current population, the set of selective values before learning is different to the set of selective values following learning:

$$\left\{s\left(e_g\right)\right\} \neq \left\{s\left(E_g\right)\right\} \tag{4}$$

where $s(h_g)$ is the selective value of an individual, g, with evaluation h_g, and $\left\{s(\ )\right\}$ is the set of selective values over the population.

Table 2. The possible transformational effects of rank respectful and rank transformational learning depending on the style of selection.

Selection Type	Rank Respectful Learning	Rank Transformational Learning
PBS	s-transforming	e-transforming; s-transforming
RBS	No effect	e-transforming

Rank Respectful Learning. The simplest models of learning may offer the same benefit to each individual in the population, or scale the evaluation of each individual by a constant factor. RBS will be unaffected by either of these forms of learning; PBS methods are unchanged by uniform scaled improvements in evaluation, but the resulting selection pressures *are* affected by uniform additive benefits applied to each individual.

For example, for the case of simple evaluation proportionate selection, where rank order is respected (i.e. there is no e-transformation) the selection pressure *against* an individual is *reduced* if:

$$\frac{e_g + b_g}{\bar{e} + \bar{b}} > \frac{e_g}{\bar{e}} \tag{5a}$$

and hence if:

$$\frac{b_g}{e_g} > \frac{\bar{b}}{\bar{e}} \tag{5b}$$

where $\bar{e}$ and $\bar{b}$ represent the population mean direct evaluation and learning benefit respectively.

In the stricter case of uniform benefits to all individuals, where:

$$b_g = \bar{b} \tag{5c}$$

equation (5b) reduces to:

$$\bar{e} > e_g \tag{5d}$$

If the *lowest* evaluation in the plasticity neighbourhood is presented to the selection function (the 'fault induction', as opposed to learning, model) then the selection pressure *against* an individual is *increased* if the inequality (5b) holds, where b_i represents the evaluation *penalty* incurred due to plasticity, and (4a) is rewritten as:

$$\frac{e_i - b_i}{\bar{e} - \bar{b}} < \frac{e_i}{\bar{e}} \tag{6}$$

Rank Transformational Learning. Most interesting forms of learing will be rank transformational. As in the rank respectful case, under PBS the selection pressures applied to individuals (and hence the selection function itself) are transformed through learning.

In both PBS and RBS schemes, where the relative ranking of individuals may be transformed by the learning operator, the selection function is essentially being passed results from a transformed *valuation* function. Strictly speaking, under both rank transformational and rank respectful learning, the selection function itself may not be transformed at all.

3.3 By Way of Example

I shall now consider surface transformation in rather more detail with respect to the MaxInt and MaxOnes evaluation functions. The learning operator I shall consider is single bit SA learning.

Firstly, we see how each valuation surface (the performance surface over the genetic representation space) is transformed through SA learning, and its complementary operator, fault induction through Steepest *Descent* 'learning', (figure 5). Note that although the same bit level operation is being applied in each case, the development function that realises assessed individuals determines whether or not the learning operator is rank respectful or rank transformational. So for example, for the case of MaxOnes, figure 5a, the valuation surface is translated and rank order is preserved (apart from 'outliers' where boundary conditions prevent learning); whereas in the MaxInt case, figure 5b, the valuation surface is significantly transformed and the original rank order is lost.

A further point to note is that MaxInt represents a one-one map from genotype to phenotype/valuation/evaluation, whereas in MaxOnes, the map is many-one. This may have consequences for IAC, (see Sect. 3.5 below).

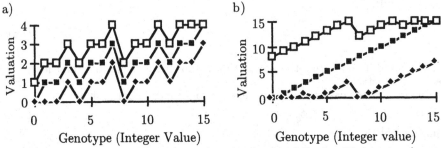

Figure 5. Direct valuation (solid square) and valuation following SA learning (empty square) and SD fault induction (solid diamond) versus integer phenotype for 4 bit MaxOnes (a) and MaxInt (b).

For the 4 bit binary coded MaxInt case, there are sixteen distinct individuals each with a unique evaluation. By introducing single bit, SA learning, the valuation profile of the population is transformed and there are only 8 distinct valuation scores, although there are still 16 distinct genotypes. What learning is doing in this case is masking genetic diversity from the selection function with valuation equivalence. In the next section, I shall demonstrate how an additional transformation of the *population structure* occurs when IAC is supported.

Example 3.1. For the MaxOnes and MaxInt evaluation functions, application of the single bit SA and SD local search operators to the population of Example 1.1, (1, 3, 5, 6, 6, 7, 8, 9, 11, 13), transforms the valuation surfaces over the population to those of figure 6 and table 3.

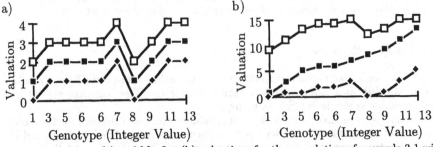

Figure 6. MaxOnes (a) and MaxInt (b) valuations for the population of example 3.1 with no learning, SA learning and SD fault induction.

Table 3. A population of 4 bit binary coded individuals given by the population set $(1, 3, 5, 6, 6, 7, 8, 9, 11, 13)$ under single bit SA learning and SD fault induction.

Naive Binary	Naive Int/Bitcount		'Educated' Binary/Int/Bitcount			'Lower bound' Binary/Int/Bitcount		
0001	1	1	1001	9	2	0000	0	0
0011	3	2	1011	11	3	0001	1	1
0101	5	2	1101	13	3	0001	1	1
0110	6	2	1110	14	3	0010	2	1
0110	6	2	1110	14	3	0010	2	1
0111	7	3	1111	15	4	0011	3	2
1000	8	1	1100	12	2	0000	0	0
1001	9	2	1101	13	3	0001	1	1
1011	11	3	1111	15	4	0011	3	2
1101	13	3	1111	15	4	0101	5	2

3.4 Cheating the Reaper - The Interaction of Learning and Selection

In the previous section, I demonstrated how the introduction of learning operator may transform the valuation landscape over the genetic representation space. As I shall now show, the choice of selection function determines the extent of further surface transformation.

Example 3.2. Using the population of Example 3.1 and Table 3, 4 bit MaxOnes and binary coded MaxInt evaluation functions, figure 7 shows the selective surfaces over the population transformed by single bit SA and SD operators under proportional selection. Figure 8 shows the selective surfaces over the MaxInt population transformed through SA learning under the truncation selection of Example 1.1.

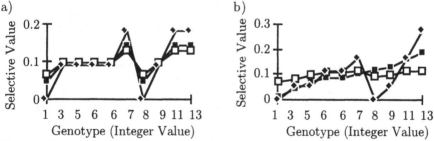

Figure 7. Proportional selective surfaces for the example population under SA learning and SD fault induction; a) MaxOnes; b) MaxInt.

From figure 7, under proportional selection, 'learning the leveler' narrows the range of selection pressures applied against population members. In RBS, figure 8, whilst the range of selection pressures remains the same, 'learning the rank transformational leveler' masks genetic diversity with phenotypic equivalence. Where rank is respected, as in MaxOnes for the particular learning operator used here, there is no transformation of the selective surface.

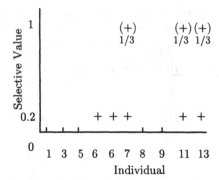

Figure 8. Truncation selective surface for the example population under SA learning for the 4 bit MaxInt evaluation function.

3.5 Transforming Population Structure Through Inheritance

I have already alluded to the transformation of population structure that results from the introduction of IAC. Where IAC is supported, the genotype corresponding to the phenotype acquired through learning is passed to the selected population, rather than the selected individual's inherited genotype. This has the consequence of transforming the population's genetic variance, as well as the evaluation received by each individual.

Recall how valuations on the 4 bit binary coded MaxInt function were transformed by single bit SA learning (figure 5b) so that only 8 distinct valuations were possible. Under IAC, the genotypes representing these 8 values are the only ones that may be passed to the breeding population. IAC, then, potentially restricts the amount of genetic variation in the breeding population.

Reaping the Benefit - The Interaction of IAC and Selection. Having transformed the population structure through IAC, I now consider how the surface of selective value is affected.

Example 3.3 Just to tie up the ends, I complete the discussion of the simple 4 bit MaxInt problem by looking at the action of the previously specified truncation selection function on the original example population as transformed by the inheritance of SA acquired characteristics. See figure 9 below. Note that the genetic variation within the sub-population up for selection is limited to two distinct individuals - 14 and 15.

4 So What's the Damage?

I have shown how the introduction of learning may give rise to valuation equivalence of genetically distinct individuals. Where such equivalence resides, there is no way of using selection to distinguish between individuals that apply learning to acquire a particular trait and those that directly inherit it, and this may lead to the introduction of a significant amount of 'selective noise' particularly in RBS. It would thus be useful to be able to distinguish on grounds of valuation between individuals that inherit a trait and those that must acquire it. A biological example is the case of the Baldwin Effect for which the necessity of a learning cost has been demonstrated by [19] - the cost

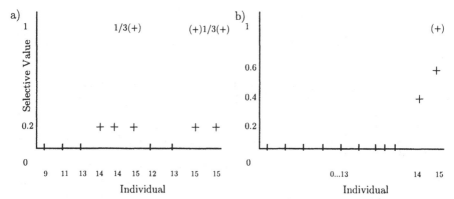

Figure 9. 4 bit MaxInt with SA learning and IAC: a) surface of selective value over the example population; b) surface of selective value for the effective population.

gives rise to the selection pressure necessary for the plastic trait to be driven out in favour of the appropriate fixed trait. More generally, where the intention is to apply an operator under evolutionary (selective) control, a cost should be associated with that operator so as to allow the selection function to distinguish between individuals that make use of the operator and those that do not, all other things being equal. The operator will then only be applied where it is selectively advantageous and as such may be usefully placed under evolutionary control.

By introducing an operator cost, such as a cost for learning, the valuation landscape which has already been transformed through the application of the learning operator, is further transformed by the exacted learning cost, figure 10. The costed valuation, $E_{g,c}$ is now given by:

$$E_{g,c} = e_g + b_g - c_g \tag{7}$$

where c_g represents the cost of learning, and $E_{g,c}$ replaces E_g in (3) and (4).

That is, the learning cost equals the difference in valuation between an individual that inherits a trait and a costed individual who has to learn it.

By considering the graph of expected evaluation for H&N (figure 10, based on [19]) it is easy to see how there is an implicit learning cost associated with learning in that experiment arising from the form of the valuation function.

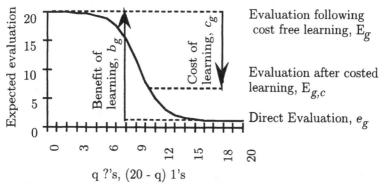

Figure 10. The cost of learning in H&N (based on [19]).

5 Rugged Mountains or Rolling Hills?

A clarification of the notion of fitness has been offered by distinguishing between the notions of valuation, evaluation and selective value. The possible search spaces underlying the corresponding surfaces, or landscapes, have been discussed and the effect of a developmental map has shown how the *evaluation* of an individual defined over a phenotypic space may be transformed into *valuation* surfaces with varying degrees of ruggedness when considered over the genotypic space. Selection acts on individual valuations in the context of the current population to determine the selective value of an individual.

I have discussed graphically the transformation of valuation landscapes by adaptive and maladaptive local search and the consequent effect on the selective surfaces induced by the particular selective conditions in use. Typically in PBS, adaptive local search (learning) smoothes the valuation and selective landscapes, whereas maladaptive local search (fault induction) sharpens them. A distinction has been made between two forms of selection and learning - proportional and rank based selection on the one hand, rank respectful and rank transformational learning on the other. These combine separately to produce two marked effects. First, there is the possibility that the range of selection pressures over the population is reduced (proportional selection, rank respectful learning); the limiting case of this is when genetic variance is masked through genetically different individuals acquiring similarly evaluated traits. Secondly, congenitally weak individuals may be hidden from the selection function at the expense of individuals with a higher direct valuation (rank transformational learning, either form of selection). The transformation induced by selection (particularly the truncation selection function considered herein) may also mask the range of direct objective values pertaining to population members, at least in so far as the differential selection of independent, 'fit' members goes.

When the inheritance of acquired characteristics is supported, not only is there a potential loss of phenotypic variation within the population, but also a corresponding loss in genotypic diversity.

Finally, I have shown how further valuation surface transformations may be forced by associating a cost with an operator, such that it may be placed under effective evolutionary control.

Acknowledgments

Thanks to the participants of the AISB EC workshop, an anonymous reviewer, Jon Rowe and Russell Anderson for their comments on earlier versions of this paper.

References

1. Wright, S. (1988) "Surfaces of Selective Value Revisited." *The American Naturalist* 131(1):115-123.
2. Provine, WB. (1986) *Sewall Wright and Evolutionary Biology.* University of Chicago Press.
3. Goldberg, DE. (1989) *Genetic Algorithms in Search, Optimization & Machine Learning.* Addison Wesley.
4. Byerly, HC, & Michod, RE. (1991) "Fitness and Evolutionary Explanation." *Biology and Philosophy* 6:1-22.
5. Sober, E. (1984) *The Nature of Selection.* Bradford Books.
6. Manderick, B, de Weger, M, & Spiessens, P. (1991) The Genetic Algorithm and the Structure of the Fitness Landscape. In *Proceedings of the Fourth International*

Conference on Genetic Algorithms (ICGA 4), pp 143-150. Eds. RK Belew & LB Booker. Morgan Kaufmann.

7. Hirst, AJ. (1996) "Search Space Neighbourhoods as an Illustrative Device." In Proceedings of *WSC1*, pp. 49-54. Nagoya University.

8. Miller, BL, & Goldberg, DE. (1996) Genetic Algorithms, Selection Schemes, and the Varying Effects of Noise. Evolutionary Computation 4(2): 113-131.

9. Nix, AE, & Vose, MD. (1991) "Modeling Genetic Algorithms with Markov Chains". Annals of Mathematics and Artificial Intelligence 5: 79-88.

10. Culberson, JC. (1994) "Mutation-Crossover Isomorphisms and the Construction of Discriminating Functions." *Evolutionary Computation* 2(3): 279-311.

11. Gitchoff, P, & Wagner, G. (1996) "Recombination Induced HyperGraphs: A New Approach to Mutation-Recombination Isomorphism." *Complxity* 2: 37-43.

12. Jones, T. (1994) A Model of Landscapes. Sante Fe Institute, Report SFI TR 95-02-21. February 1, 1994.

13. Wagner, GP, & Altenberg, L. (1996) "Perspective - Complex Adaptations and the Evolution of Evolvability." Evolution 50(3): 967-976.

14. A elmeyer, T, Ebeling, W, & Rosffi, H. (1996) "Smoothing Representation of Fitness Landscapes - The Genotype-Phenotype Map of Evolution." *Biosystems* 39(1): 63-76.

15. Radcliffe, NJ, & Surry, PD. (1994) "Formal Memetic Algorithms." In *Evolutionary Computing: AISB Workshop*. Ed. T Fogarty. Springer Verlag.

16. Whitley, D, Gordon, VS, & Mathias, K. (1994) "Lamarckian Evolution, The Baldwin Effect and Function Optimization." In Proceedings of *Parallel Problem Solving Fron Nature PPSN III*, 6-15. Eds. Y Davidor, HP Schwefel & R Manner. Springer-Verlag.

17. Hinton, GE, & Nowlan, SJ. (1987) "How Learning Can Guide Evolution." *Complex Systems* 1:497-502.

18. Harvey, I. (1993) "The Puzzle of the Persistent Question Marks: A Case for Genetic Drift." In *Genetic Algorithms: Proceedings of the 5th International Conference*, 15-22. Ed. S Forrest. Morgan Kaufmann.

19. Mayley, G. (1996) No Pain, No Gain: Landscapes, Learning Costs and Genetic Assimilation. CSRP 409, COGS, Sussex University. February, 1996.

Island Model Genetic Algorithms and Linearly Separable Problems

Darrell Whitley, Soraya Rana, and Robert B. Heckendorn

Department of Computer Science
Colorado State University
Fort Collins, Colorado 80523 USA

Abstract. Parallel Genetic Algorithms have often been reported to yield better performance than Genetic Algorithms which use a single large panmictic population. In the case of the Island Model Genetic Algorithm, it has been informally argued that having multiple subpopulations helps to preserve genetic diversity, since each island can potentially follow a different search trajectory through the search space. It is also possible that since linearly separable problems are often used to test Genetic Algorithms, that Island models may simply be particularly well suited to exploiting the separable nature of the test problems. We explore this possibility by using the infinite population models of simple genetic algorithms to study how Island Models can track multiple search trajectories. We also introduce a simple model for better understanding when Island Model Genetic Algorithms may have an advantage when processing linearly separable problems.

1 Introduction

Island Models are a popular and efficient way to implement a genetic algorithm on both serial and parallel machines[1, 12, 20, 7]. In a parallel implementation of an **Island Model** each machine maintains its own subpopulation using a genetic algorithm for search. The machines work in consort by periodically exchanging a portion of their populations in a process called **migration**. For example, a total population N_{total} for a serial algorithm could be spread across M machines by giving each machine a population size of $N_{island} = N_{total}/M$. The Island Model introduces the two parameters **migration interval**, the number of generations (or evaluations) between a migration, and **migration size**, the number of individuals in the population to migrate.

Parallel Island Models have often been reported to display better search performance than serial single population models, both in terms of the quality of the solution found as well as reduced effort as measured in terms of total number of evaluations corresponding to points sampled in the search space [9, 20]. One reason for the improved search quality is that the various "islands" maintain some degree of independence and thus explore different regions of the search space while at the same time sharing information by means of migration. This can also be seen as a means of sustaining genetic diversity [11]. Some researchers [9, 2] have gone back to the work of Fisher [4] and Wright [21] in biology to try to better understand the role of locality (e.g., maintaining distinct islands or some other form of spatial separation) in evolution.

The partially isolated nature of the island populations suggests that Island Models may be well adapted for use on problems that are loosely composed of many separate problems that can be solved independently. Many test problems that have been used to measure the performance of genetic algorithms turn out to have exactly this property in that they are **linearly separable**. That is, the problem can be decomposed into the sum of a set of subproblems each of which can be solved independently and each of which usually takes a limited subset of the entire set of function arguments.

Island model genetic algorithms have sometimes done well against single population models on such test problems [11, 9]. We examine whether Island Model Genetic Algorithms, in fact, do take advantage of the linear separability of a problem to improve performance by solving each subproblem separately in island populations and assembling the partial solutions into a complete solution through the use of migration.

We begin by reviewing the evidence which suggests that having multiple islands can generally improve search performance. In particular, we show that islands can exploit separate and distinct fixed points in the space of possible populations for a single population genetic algorithm. We then propose a simplified probabilistic model of Island Model Genetic Algorithms to further motivate our explorations. We present the test problems and algorithms we used and discuss our results. These results suggest that our hypothesis of improved performance for the Island Model on linearly separable problems is only partly correct. Finally, we discuss possible mechanisms that might explain our observations and explore the pros and cons of the Island Model Genetic Algorithm.

2 Initial Conditions and Island Model GAs

Vose [13] and Whitley et al. [14] independently introduced exact models of a simple genetic algorithm using infinite population assumptions. Whitley [19] provides an alternate, more introductory derivation of these models. Vose [10] extends this model to look at finite populations using Markov models.

One of the difficulties of using the infinite population model is that one cannot take into account the sampling bias introduced by using finite populations. The finite population Markov model on the other hand, is too expensive to actually execute except for extremely small problems and extremely small populations (e.g. [8]). One way in which we can introduce finite effects into the infinite population model is by initializing the infinite population model using a distribution taken from a finite population.

In this section, we present a very simple application of the infinite population model to look at the behavior of multiple simple genetic algorithms running in parallel which also exchange information using an idealized form of migration. The multiple simple genetic algorithms have different behaviors due to the different initial conditions that result from initializing the models using different finite population samples.

In particular, we were interested to see if we could find behaviors consistent with the claims that have been made regarding parallel island genetic algorithms: we wanted to see if different islands following different search trajectories could be a significant source of genetic diversity. Mutation is not used in these experiments, since we wanted to better isolate genetic diversity resulting from the interaction between islands.

2.1 The Infinite Population Model

Vose [13] uses the function $\mathcal{G}$ to represent the trajectory of the infinite population simple genetic algorithm [6]. The function $\mathcal{G}$ acts on a vector, p, where component p_i of the vector is the proportion of the population that samples string i. We denote the population at time t by p^t. Note that p^t may be the sampling distribution of a finite or an infinite population. Using function $\mathcal{G}$ we can generate

$$p^{t+1} = \mathcal{G}(p^t)$$

where p^{t+1} is the expected next generation of an infinitely large population. The vector p^{t+1} also is the expected sampling distribution utilized by any finite population. Details on the construction of $\mathcal{G}$ are given by Whitley [19]; the models have also been extended to include Hybrid Genetic Algorithms using local search as a supplemental search operator [15].

The function $\mathcal{G}$ exactly corresponds to the behavior of a simple genetic algorithm when the population is infinitely large. In the short term, the trajectory of $\mathcal{G}$ moves to a fixed point in the space of possible populations; in the long term, it may transition to other fixed points due to the effects of mutation. When working with finite populations, larger populations tend to track the behavior of $\mathcal{G}$ with greater fidelity than smaller populations.

Populations can diverge from the trajectory of $\mathcal{G}$ if the actual trajectory followed by the finite population crosses over into the basin of attraction for a different fixed point. Also, due to the effects of sampling, the *initial population* p^0 of a finite population can sample only a small finite subset of the 2^L strings of length L that make up the search space. Thus, finite populations may actually start in a different basin of attraction than an infinitely large population that uniformly samples all 2^L strings in the space.

This immediately suggests that there may be a potential advantage in running several smaller populations for a shorter amount of time instead of running a single larger population for a longer amount of time. In other words, the fact that small populations are less likely to follow the trajectory of the infinite population model could be an advantage because the different subpopulations potentially move into "different basins of attraction", and thus, potentially explore different parts of the search space. It should be stressed that these "different basins of attraction" exist with respect to the single population model. An island model induces a different space of possible population configurations and of course can only be in one configuration at a time.

While the function G assumes an infinitely large population, we can use G to examine the impact of starting the single population genetic algorithm or the island models from different initial population distributions based on actual finite population samples.

We can use G in a simple way to model an idealized Island Model Genetic Algorithm. The island model assumes that the total population used by the genetic algorithm is broken down into a number of subpopulations referred to as islands. Each island is in effect executing a version of the genetic algorithm on each subpopulation. In the model used here, each island is executing a separate copy of G for the standard simple genetic algorithms using 1-point crossover. *Migration* occurs between subpopulations in a restrictive fashion based on temporal and spatial considerations. The crossover rate is 0.6.

In the island model used here, migration occurs every X generations and copies of the individuals that make up the most fit 5% of the island population are allowed to migrate. This top 5% may be copies of the same string, or it may be some combination of strings. The island receiving these strings deletes the least fit bottom 5% of its own population. The migration scheme assumes that the islands are arranged in a ring. On the first migration, strings move from their current island to their immediate neighbor to the left. Migrations occur between all islands simultaneously. On the second migration, the islands send copies of the top 5% of their current population to the island which is two moves to the left in the ring. In general, the migration destination address is incremented by 1 and moves around the ring. Migrations occur every X generations until each island has sent one set of strings to every other island (not including itself); then the process is repeated.

2.2 Modeling Island Trajectories

Four subpopulations are modeled. The distribution of strings in the subpopulations are initialized by generating distinct finite subpopulations. The initial distribution for each island was based on a sample of 80 random strings ($80 = 2^4 * 5$). However, infinite population sizes are still assumed after the initial generation. Thus, the trajectories only model the effects of having different starting points in the different subpopulations.

Results for a 4 bit fully deceptive problem [18] are shown in Figs. 1 and 2 with a global optimum of 1111 and a local optimum of 0000. The exact function is also given in section 3.2 of this paper. The choice of a fully deceptive function is in no way critical to illustrate the computational behavior of the models. It is important, however, to use a function with competing solutions that are some distance removed from one another in Hamming Space and that the GA-surface induced by G for a single population GA have multiple fixed points. On functions with a single optimum, the GA produces predictable behavior: quick monotonic convergence to the global optimum in all subpopulations.

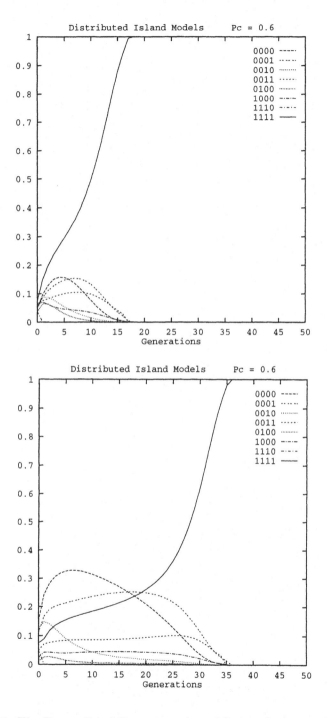

Fig. 1. The Proportional Representation of Strings in 2 out of 4 Islands When Migration Occurs **Every** Generation.

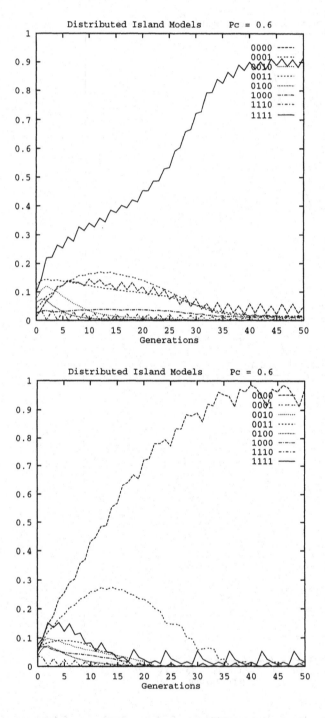

Fig. 2. The Proportional Representation of Strings in 2 out of 4 Islands When Migration Occurs **Every Second** Generation.

The graphs in Fig. 1 show the computation of two out of four islands when migration occurs every generation. In this case, the islands all show the same convergence behavior, although the islands vary in the low level details of their behavior. The other two islands (not shown) had behavior that was quite similar to one of these two. Thus, two of the islands quickly converge to the global optimum, while two of the islands initially moved toward the local optimum at 0000 and are later drawn toward 1111, the global optimum.

It should be noted that a single population model started with a uniform representation of strings converges to 0000. One reason that the island models converge to 1111 is that migration of the best strings acts as an additional selection mechanism. The best strings are not only increasing due to fitness proportionate selection; the strings that migrate also tend to double their representation, while the worst strings are deleted.

The graphs in Fig. 2 show two out of four islands when migration is occurring every *second* generation. Migration produces the spike-like jumps in representation, since local representations will rise or fall when migration occurs, but migration does not occur at each time step. These jumps in representation did not show up when migration occurred every generation since its effect is averaged in at each time step. Jumps in representation always occur, however, when the time between migrations is greater than one generation.

Figures 1 and 2 show that the migration of the best strings acts to increase selection due to fitness. Convergence is faster in the graphs in Fig. 1 where migration occurs every generation. Furthermore, the additional selective pressure produced by migration helps to drive the convergence to the global optimum in the topmost graphs. With migration every generation, all subpopulations converge to 1111. With migration only every second generation, only one of the four subpopulations converge to 1111 and convergence is more gradual. (The results for the two islands not shown are similar to the computation represented by the bottom graph.) In additional runs, not shown here, migrating every 3 generations resulted in all subpopulations converging to the local optimum 0000 (just as in the single population case when there is no additional selection boost due to migration) and again the convergence rate is slower.

The results presented in the figures also suggest that the island model may be especially well suited to handling the additional selective pressure. Although the island model equations are restricted by the assumption of infinite populations *after the first generation*, the models nevertheless show that having different starting points for the individual islands may be sufficient to cause each island to follow a unique search trajectory. And although migration will tend to cause the set of equations representing the various islands to become more similar over time, surprisingly, the results using migration every second generation show that it is possible for different islands to converge to different solutions. When these results are extended forward in time, the pattern emerging in the graphs in Figure 2 around generation 35 appears to be stable.

These results also show how convergence to distinct populations that would be near different fixed points in the infinite single population case actually can help to preserve genetic diversity. Normally, without a mutation operator a population would converge to a single string and thus become stuck. However, with the two islands converging to different solutions and with migration to do low level mixing, new strings are constantly generated at a low level.

3 Linearly Separable Problems

In the previous section, a general argument was presented to explain why island model genetic algorithms can display improved search performance compared to single population models. In this section, we specifically look at the application of island models to decomposable separable functions.

3.1 A Probabilistic Argument for Better Performance

A probabilistic argument can be made for why an Island Model Genetic Algorithm may display more rapid convergence to a given quality of answer for a linearly separable problem. Consider an evaluation function F composed of the sum of S independent nonlinear subproblems G_i:

$$F(V_1, V_2, ...V_S) = G_1(V_1) + G_2(V_2) + ... + G_S(V_S).$$

Assume we apply a genetic algorithm with population size N_x, which solves a specific subproblem G_i from F in t evaluations with probability X_t. Similarly, a genetic algorithm with population N_y solves the same subproblem in t evaluations with probability Y_t. Let $N_y = M * N_x$. In this case N_x represents an island population size for M processors giving a total population size of N_y.

For simplicity, we make several strong assumptions. First, we assume all the subfunctions G_i are identical. We assume that each subproblem G_i is solved *independently* during genetic search, regardless of whether a single population or multiple subpopulations are used. We assume that migration and recombination between subpopulations will build a solution in a post-hoc fashion from the best solution for each subproblem out of the various subpopulations. And finally, we assume the amount of time it takes to assemble to full solution is small in comparison with the time to find the individual solutions.

A subproblem G_i is **solved** when the correct value is found to optimize G_i, or alternatively, when a solution within ϵ of the optimal solution is found. The exact conditions for a problem being solved in our empirical tests are specified for each problem in our test suite.

Given these assumptions we may reason as follows. Consider an Island Model Genetic Algorithm with M islands. Let $X_{t/M}$ be the probability of any one island solving a specific G_i in t/M evaluations and Y_t be the probability of the total population in a single population model solving the same G_i. Note that the probabilities $X_{t/M}$ and Y_t are the same for all G_i under our assumptions. It is clear that the M island model and the single population model will use t total evaluations but the probability of solution may be different. It is this difference that may justify our expectation of a performance improvement.

Since we have M subpopulations, we have M chances of solving the subproblem G_i. Under our model, G_i is solved if it is solved in any subpopulation. The probability that G_i is **not** solved by **any** island population is given by

$$(1 - X_{t/M})^M$$

therefore the probability that it is solved is

$$1 - (1 - X_{t/M})^M$$

Since there are S subproblems with identical probabilities to solve and we assumed they are solved independently, the probability of solving all S is

$$(1 - (1 - X_{t/M})^M)^S$$

In the case of a single population, any one of the subproblems will be solved with a probability Y_t, hence it solves all S independent problems with probability $(Y_t)^S$.

Under what conditions would we expect the probability to be greater by using islands than using a single population? The reasoning above gives us an upper bound on Y_t for this condition:

$$1 - (1 - X_{t/M})^M > Y_t$$

From fixed values of $X_{t/M}$ and Y_t an estimate of the number of islands necessary for equivalent performance can be made as follows. If we assume

$$1 - (1 - X_{t/M})^{M'} = Y_t$$

we can solve for M'

$$\frac{\ln(1 - Y_t)}{\ln(1 - X_{t/M})} = M'$$

which is the number of islands, M', of population N_{total}/M running for t/M evaluations each that would be necessary to have the same likelihood of solving F as a single population of size N_{total} running for t evaluations.

For example, if we know the probabilities $X_{t/M} = 0.05$ and $Y_t = 0.1$ for $M = 10$ and populations of size $N_{island} = 200$, $N_{total} = 2000$ then the model predicts that with $M' = 2$ subpopulations of 200 each running for $t/10$ evaluations will yield approximately the same results as a single population of 2000 running for t evaluations. That is:

$$(1 - (1 - X_{t/M})^{M'}) \approx Y_t$$
$$(1 - (1 - 0.05)^2)^S \approx .1^S$$

In cases where $M' \leq M$ we can conclude that increasing the number of machines from M' to M will only improve the odds of finding the solution. In this case, an Island Model Genetic Algorithm with 10 subpopulations of 200 each (i.e., a total population of 2000) will probably yield much better performance than just two subpopulations of 200 or a single population of 2000. In general, when the ratio $\ln(1 - Y_t)/\ln(1 - X_{t/M})$ is less than M then the Island Model may display a performance advantage over a single population.

For example, if $X_{t/M} = 0.01$ and $Y_t = 0.1$ when M $= 10$, the single population will have a slight advantage. (Note that for the range of small probability values we have used in these examples, the ratio between $X_{t/M}$ and Y_t directly tracks the linear scaling of the number of subpopulations.) In reality, the 10 subpopulations will probably be at a real disadvantage because the various distributed subsolutions have to be migrated and assembled, and this process is not trivial.

It should be noted that this model also generalizes to cover nonseparable problems. In this case, however, we must know the probability that the entire problem is solved as opposed to subproblems. In other words, the model also covers the special case where we have a single subproblem.

There are obviously also many effects that are ignored by this model. For example, **Hitch-hiking** occurs when strings that carry bit values that solve one subproblem, say G_i, carry bit values elsewhere in the string that do not solve other subproblems G_k, $i \neq k$. As a string propagates due to selection, it also propagates other bit values that do not solve G_k. Overall however, the model makes the point that for decomposable, separable problems, that if the ability to solve multiple subproblems does not scale with population size, then there may be an advantage to using several small populations rather than one large single population.

3.2 The Test Problems

The following two linearly separable problems were chosen as test problems because they represent examples from well known classes of linearly separable test problems. Fortunately they also display different behavior under our tests and lead to a better understanding of what factors contribute to Island Model performance.

The first is based on deceptive functions. The following is the **fully deceptive** order-4 function used in the examples presented in Figures 1 and 2.

$$
\begin{array}{ll}
\texttt{f(1111) = 30} & \texttt{f(0000) = 28} \\
\texttt{f(0111) = 0} & \texttt{f(1011) = 2} \\
\texttt{f(1101) = 4} & \texttt{f(1110) = 6} \\
\texttt{f(1100) = 8} & \texttt{f(1010) = 10} \\
\texttt{f(1001) = 12} & \texttt{f(0110) = 14} \\
\texttt{f(0101) = 16} & \texttt{f(0011) = 18} \\
\texttt{f(1000) = 20} & \texttt{f(0100) = 22} \\
\texttt{f(0010) = 24} & \texttt{f(0001) = 26} \\
\end{array}
$$

On a fully deceptive function, every hyperplane competition is misleading. For example, for a 3 bit function this implies the following hyperplane fitness relationships:

```
f(0**) > f(1**)        f(00*) > f(11*),  f(01*),  f(10*)
f(*0*) > f(*1*)        f(0*0) > f(1*1),  f(0*1),  f(1*0)
f(**0) > f(**1)        f(*00) > f(*11),  f(*01),  f(*10)
```

where f(H) is the average fitness of the strings in the hyperplane denoted by H. The function is fully deceptive because the global optimum is at 111 while all hyperplanes support 000 as a potential solution. Fully deceptive functions are clearly artificial constructs; yet it would also seem unlikely for a function to have no deception. What is important for the current study is that there is strong competition between two complementary points in the space, 000 and 111, or in the case of our 4 bit fully deceptive function, 0000 and 1111. As already seen in the experiments using infinite population models, a genetic algorithm can converge toward 0000 in some circumstances and toward 1111 in other situations. This makes this an interesting test function for island models.

Goldberg, Korb and Deb [5] defined a 30 bit function constructed from 10 copies of a fully deceptive 3-bit function. The problem is made more difficult when the bits are arranged so that each of the 3 bits of the subfunctions are uniformly and maximally distributed across the encoding. Thus each 3 bit sub-function, i, has bits located at positions i, $i + 10$, and $i + 20$.

For our experiments we constructed a 40 bit function composed of 10 copies of our 4-bit fully deceptive function with the 4 bits of the 10 subfunctions distributed at positions i, $i + 10$, $i + 20$, $i + 30$, for subfunctions $i = 1$ to 10. We also picked this problem because parallel island models have previously been shown to have better results than single population approaches [11] on this problem.

The second linearly separable test function is Rastrigin's Function.

$$f(x_i \mid_{i=1,N}) = 10N + \sum_{i=1}^{N} \left(x_i^2 - 10 \cos(2\pi x_i) \right); \qquad x_i \in [-5.12, 5.11]$$

We used $N = 10$ and 10 bits per argument giving us a 100 bit representation. The bits of each argument are contiguous.

The encoding for both test problems was binary. That is, all bits are treated uniformly and no information about parameter boundaries was used in crossover and mutation. In the case of Rastrigin's function the encoding is linear scaling from the binary representation without any real number encoding such as a separate mantissa.

4   Experimental Results

The search engine in our experiments is GENITOR [17], a steady-state genetic algorithm. Each problem was run with a fixed total population size of 5000 but with varying numbers of islands as shown in Table 1. The **migration interval** in the table is the number of evaluations *per island* between migrations. The parameters in Table 1 were chosen to be the same as those used in a previous

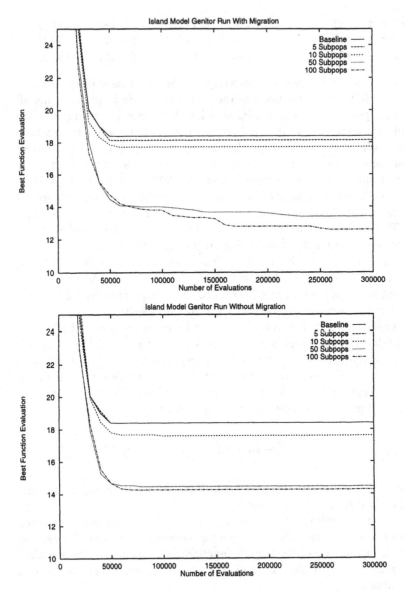

Fig. 3. The Deceptive Function with and without Migration.

study by Starkweather, Whitley, and Mathias [11]. 30 trials were run for each parameter set. Each trial was run until a solution was found or until a total of 300,000 function evaluations was reached for that trial. No mutation was used; the crossover rate was 1.0 since this is a steady-state genetic algorithm.

In these experiments, we consider a subproblem solved if the best overall solution found by GENITOR solvest that particular subproblem.

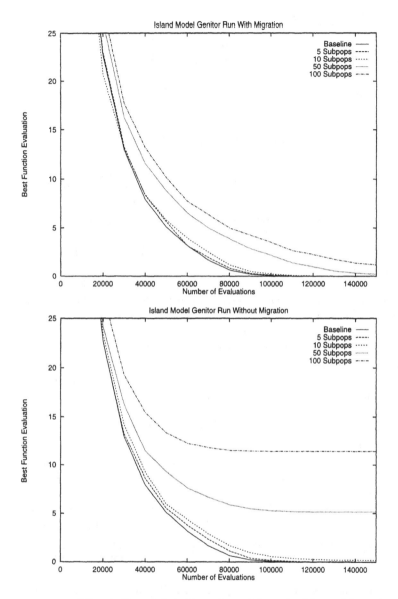

Fig. 4. Rastrigin's Function with and without Migration.

The results in Figs. 3 and 4 show that increased parallelism helps on the deceptive problem, but not for Rastrigin's function. Note that when run without migration, the islands are really just independent runs of Genitor using different population sizes. All of the results are scaled so that the x-axis is the total number of evaluations across all subpopulations.

Larger populations do not help much on the deceptive problem but do help a great deal on Rastrigin's function. This would also partially explain why the

Table 1. Experimental Population and Migration Sizes

Island Population	Number of Islands	Number of Emigrants	Migration Interval
50	100	2	250
100	50	2	500
500	10	5	2500
1000	5	5	5000
5000	1	N/A	N/A

Island Model helps on the deceptive problem: it is less sensitive to population size. This is consistent with the insights associated with our abstract model of the Island Model Genetic Algorithm; if increasing the size of the population does not increase the probability of solving a subproblem (i.e., if Y_t is not significantly better than $X_{t/M}$) then the Island Model has an advantage. The figures also show that migration is an important factor.

We should also point out that these results are peculiar in the following way: *the answers for the various subproblems are contained in the initial population of* 5000. On Rastrigin's function using a 10 bit encoding, there is 1 chance in a 1000 of randomly generating the optimal solution to any given subproblem. This is also a typical situation for deceptive problems; the deception is usually defined over subfunctions defined over a small number of bits and these subfunctions do not interact. Hence, the "solutions" to the deceptive subfunctions are also contained in the initial population. The real question is whether these "solutions" propagate under crossover.

Population size may also be more of a factor on other separable problems than many researchers may realize. Thus, what we are looking at in Figs. 3 and 4 are how readily the subsolutions spread in the populations and how much the solutions are disrupted by crossover. Of course, the "ugly" deceptive problem is designed so that crossover is disruptive. However, by looking at the experiments involving Rastrigin's function, it can clearly be seen that the various subsolutions are distributed across the population (or subpopulations) and monotonically increase their distribution in the population over time. Distributing the total population across islands in this case only makes it harder for the solution to spread.

For Rastrigin's function one might reasonably ask, what if the total population was smaller, for example 500. Fig. 5 shows the results with and without migration for Rastrigin's function. (As one might expect, changing the population size for the deceptive problem does not change the results much compared to those shown in Figs. 1 and 2 since the probably of randomly generating a solution to a deceptive subproblem is 1/16 in this case.)

The difference between the results with and without migration is now even more dramatic. The solutions are now not always contained in the initial population and now the parallel island models run with migration out perform the single population in the middle and later stages of the search.

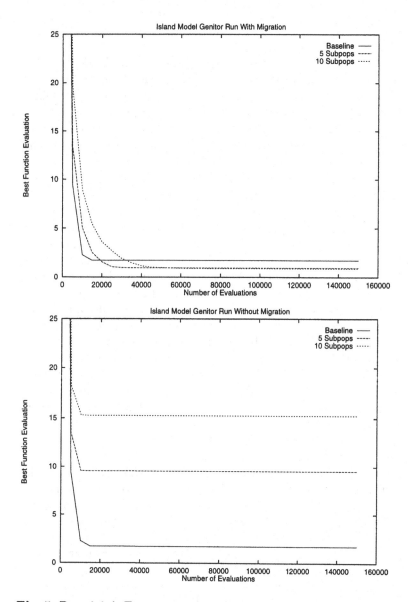

Fig. 5. Rastrigin's Function with and without Migration Using a Total Population of 500.

5 Discussion

We present an abstract model of when we might expect Island Model Genetic Algorithms to out-perform single population models on linearly separable problems. Our model suggests that Island Models may be at an advantage when increasing the population size does not help to solve the problem. There are many factors that our model does not consider however. Our empirical results suggest that Island Model Genetic Algorithms can still display very complex and unexpected behavior even when applied to separable problems.

We are well aware of the limitations of using separable problems for testing optimization search engines and have argued for more diverse types of test problems with different degrees of nonlinearity [16]. Island Model Genetic Algorithms typically use larger total population sizes; 10 subpopulations of size 50 would not be an unusual or unreasonable design. Yet, for separable functions larger overall population sizes will mean that there is an increased chance of the optimal solution being in the initial population–which can produce anomalous experimental results. On the other hand, separable functions are a special subclass of problems and studying how they are processed in island models may still prove useful despite anomalies. In particular, separable functions are nice for studying migration because the subproblems are independent.

We have run additional experiments that vary the migration rate. Conventional wisdom suggests that as the migration rate is increased the Island Model's performance is increasingly like the single population model. Our results suggest that this is true to some degree, but that the situation is still more complex than such a general "rule of thumb" might lead one to believe.

Finally, the difficulty with experimental research is that the results depend on so many parameters. Different migration patterns or migration rates might lead to different results. In our experiments, we used GENITOR for the basic search engine. Based on our abstract model, we conjecture that island models will have a better advantage against single population models if the search engine is one that works particularly well with small population sizes. This may be true regardless of whether the target problems are separable or not. Thus we expect that an algorithm such as CHC [3] would be a better choice. Scaling up the population (which is usually set to 50 for CHC) rarely improves performance for CHC, or improves performance only modestly. Thus, it should be a very good candidate for a search engine for use in Island Model Genetic Algorithms.

References

1. Theodore C. Belding. The distributed genetic algorithm revisited. In L. Eshelman, editor, *Proc. of the 6th Int'l. Conf. on GAs*, pages 114–121. Morgan Kaufmann, 1995.
2. R. Collins and D. Jefferson. Selection in Massively Parallel Genetic Algorithms. In L. Booker and R. Belew, editors, *Proceedings of the Fourth International Conference on Genetic Algorithms*, pages 249–256. Morgan Kaufmann, 1991.

3. Larry Eshelman. The CHC Adaptive Search Algorithm. How to Have Safe Search When Engaging in Nontraditional Genetic Recombination. In G. Rawlins, editor, *FOGA -1*, pages 265–283. Morgan Kaufmann, 1991.

4. R. Fisher. *The Genetical Theory of Natural Selection.* Dover, New York, 1958.

5. D. Goldberg, B. Korb, and K. Deb. Messy Genetic Algorithms: Motivation, Analysis, and First Results. *Complex Systems*, 4:415–444, 1989.

6. David Goldberg. *Genetic Algorithms in Search, Optimization and Machine Learning.* Addison-Wesley, Reading, MA, 1989.

7. Martina Gorges-Schleuter. Explicit Parallelism of Genetic Algorithms through Population Structures. In H.P. Schwefel and Reinhard Männer, editors, *Parallel Problem Solving from Nature*, pages 150–159. Springer/Verlag, 1991.

8. Ken De Jong, William Spears, and Diana Gordon. Using Markov Chains to Analyze GAFOs. In D. Whitley and M. Vose, editors, *FOGA - 3*. Morgan Kaufmann, 1995.

9. H. Mühlenbein. Evolution in Time and Space: The Parallel Genetic Algorithm. In G. Rawlins, editor, *FOGA -1*, pages 316–337. Morgan Kaufmann, 1991.

10. A. Nix and M. Vose. Modelling Genetic Algorithms with Markov Chains. *Annals of Mathematics and Artificial Intelligence*, 5:79–88, 1992.

11. Timothy Starkweather, L. Darrell Whitley, and Keith E. Mathias. Optimization Using Distributed Genetic Algorithms. In H.P. Schwefel and R. Männer, editors, *Parallel Problem Solving from Nature*, pages 176–185. Springer/Verlag, 1990.

12. Reiko Tanese. Distributed Genetic Algorithms. In J. D. Schaffer, editor, *Proceedings of the Third International Conference on Genetic Algorithms*, pages 434–439. Morgan Kaufmann, 1989.

13. M. Vose and G. Liepins. Punctuated Equilibria in Genetic Search. *Complex Systems*, 5:31.44, 1991.

14. D. Whitley, R. Das, and C. Crabb. Tracking Primary Hyperplane Competitors During Genetic Search. *Annals of Mathematics and Artificial Intelligence*, 6:367–388, 1992.

15. Darrell Whitley. Modeling Hybrid Genetic Algorithms. In G. Winter, J. Periaux, M. Galan, and P. Cuestra, editors, *Genetic Algorithms in Engineering and Computer Science*, pages 191–201. Wiley, New York, 1995.

16. Darrell Whitley, Keith Mathias, Soraya Rana, and John Dzubera. Evaluating Evolutionary Algorithms. *Artificial Intelligence Journal*, 85, August 1996.

17. L. Darrell Whitley. The GENITOR Algorithm and Selective Pressure: Why Rank Based Allocation of Reproductive Trials is Best. In J. D. Schaffer, editor, *Proceedings of the Third International Conference on Genetic Algorithms*, pages 116–121. Morgan Kaufmann, 1989.

18. L. Darrell Whitley. Fundamental Principles of Deception in Genetic Search. In G. Rawlins, editor, *FOGA -1*, pages 221–241. Morgan Kaufmann, 1991.

19. L. Darrell Whitley. A Genetic Algorithm Tutorial. *Statistics and Computing*, 4:65–85, 1994.

20. L. Darrell Whitley and Timothy Starkweather. GENITOR II: A Distributed Genetic Algorithm. *Journal of Experimental and Theoretical Artificial Intelligence*, 2:189–214, 1990.

21. Sewell Wright. The Roles of Mutation, Inbreeding, Crossbreeding, and Selection in Evolution. In *Proceedings of the Sixth International Congress on Genetics*, pages 356–366, 1932.

Evolutionary Machine Learning and Classifier Systems

Empirical Validation of the Performance of a Class of Transient Detector

Philip J. Jacob and Andrew D. Ball

School of Engineering, University of Manchester, Manchester. M13 9PL. UK.

Abstract. Transient detection in the presence of noise is a problem which occurs in many areas of engineering. A description is given of a classifier system suitable for the identification of high frequency waveforms. It uses the Wavelet Transform for signal pre-processing to produce a more parsimonious representation of the signal to be identified. A comparison is presented of the use of a Forward Selection algorithm and a Genetic Algorithm to pick appropriate indicator variables as inputs to a classifier. A Radial Basis Function neural network is employed to model the class conditional probability density function. The classifier is applied to the identification of a number of high frequency Acoustic Emission signals, which are difficult to classify,.

1. Classifiers in Fault Detection

1.1 Introduction

Transient detection in the presence of noise is a problem that occurs in many areas of engineering such as radar signature identification, seismic surveillance and fault detection. The transients may be characterised by a feature vector formed from a transform of the signal to produce a characteristic feature vector. Much recent work has been dedicated towards the development of a class of transforms called Wavelet Transforms. Wavelet Transforms have the dual advantages of allowing multi-resolution analysis of a signal, and, with a careful choice of suitable filter bank coefficients, produce a transformed vector which is fairly sparse. Time-scale transforms are useful for characterising and classifying non-stationary signals since signals with near identical spectra may have very distinct time-scale representations.

The classification of waveforms may be regarded as a three stage process:-

1. Signal processing of the sampled waveform to derive a vector representing the entire waveform. Such processing will typically involve sampling of the analogue signal, followed by transformation of the time domain data vector into the frequency domain to generate a power spectrum, or with an linear orthogonal transform such as a wavelet transform.

2. Feature extraction - selecting a subset of the elements to include in a reduced feature vector from the original feature vector which are most suitable for using as the information on which to base a classification decision. A comparison is presented of the performance of a Genetic Algorithm and a Forward Selection heuristic to choose a suitable set of features.

3. Modelling of the *posteriori* class membership probability distribution, that is the likelihood that a waveform is a member of a particular class, based on the values in its reduced feature vector.

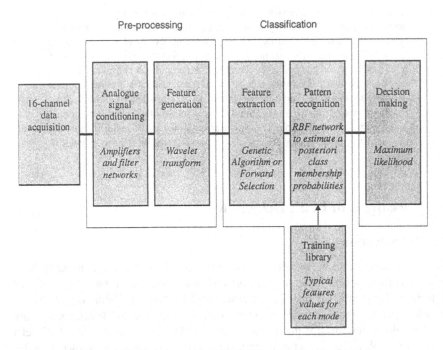

Figure 1.1- Components of a high frequency classifier system

In this paper an architecture is discussed for a classification system suitable for classifying high frequency, transient waveforms. Use is made of the Wavelet Transform to effect the signal pre-processing. A comparison is then presented of the performance of a forward selection-based algorithm and a Genetic Algorithm is to select suitable features for inclusion in the reduced feature vector. A Radial Basis Function neural network is employed to establish a non-parametric mapping between the reduced feature vectors and the classification target vector. Components of this system are illustrated in Figure 1.1.

1.2 Statistical Pattern Recognition

An outline is given of the fundamental concepts of statistical pattern recognition by considering the simple problem of identifying N classes of waveform. Each waveform is represented by a vector of length d of samples of the waveform, $\mathbf{x} = (x_1, \ldots, x_d)^T$. The aim of classification is to develop an algorithm which will assign any waveform represented by a vector such as $\mathbf{x}$ to one of N classes, which will be written C_j, where $1 \leq j \leq N$. We shall assume that we are supplied with a large number of examples of vectors corresponding to waveforms from each of the N classes and that these vectors have already been labelled by some means with the correct classification, C_k.

Unfortunately however, as the signals are derived from very high frequency sources, a large number of samples will be required to capture an event of any reasonable duration. The classifier system has at most a few thousand examples of labelled data on which it is to base its future predictions and to set the free parameters of the classifier. The problem of generalisation is very difficult in this case, since it is not possible to assign values to the free parameters within the classifier system with any reasonable degree of statistical certainty based on such a relatively small number of training examples. This necessitates the use of some form of feature extraction process to reduce the length of the training data vectors and, thereby, allow the construction of a classifier with fewer free parameters. It is to be hoped that the information contained in the reduced feature vector, $\hat{\mathbf{x}}$, will be sufficient so as to assign the vectors to their correct class. This will not necessarily be the case.

In order to calculate the *posteriori* probability density function, $P(C_j|\mathbf{x})$ where C_j indicated membership of class j, $1 \leq j \leq N$, use is made of two other probabilities: the class conditional probability distribution of $\mathbf{x}$ for class C_j, $P(\mathbf{x}|C_j)$, and the prior probability $P(C_j)$. This second prior probability term is simply the probability that any given data point in the data set belongs to class C_j, that is the number of class C_j items divided by the total number of items in the training data set. It is possible to calculate this prior for each of the N classes in the labelled data set. The class conditional probability density $P(\mathbf{x}|C_j)$ is the probability that an example of a waveform from class j could have the feature vector $\mathbf{x}$. These two probabilities are combined using Bayes' rule and normalise to give the *posteriori* probability of class membership [1],

$$P(C_j|\mathbf{x}) = \frac{P(\mathbf{x}|C_j)P(C_j)}{p(\mathbf{x})} \qquad (1.2.0)$$

The term p(x) is the unconditional density function, that is the density function of $\mathbf{x}$ irrespective of the class and is given by

$$p(\mathbf{x}) = \sum_{j=1}^{N} P(\mathbf{x}|C_j)p(\mathbf{x}) \qquad (1.2.1)$$

This normalisation factor ensures that the probabilities sum to 1. The class with the maximum *posteriori* probability of class membership is then chosen and the vector assigned to this class. The remaining difficulty lies in modelling the class conditional probability distribution $P(x|C_j)$. This will be carried out using a Radial Basis Function neural network as described in section 4.

2. Wavelet Transforms

Viewed in the time domain, the energy of the Acoustic Emission signal from an impact-type event, x, is widely distributed across all of the coefficients. The standard basis functions provide a representation which is highly unparsimonious. For a signal with D samples, these standard basis functions are:

$$\mathbf{b}_0 = \begin{bmatrix} 1 & 0 & 0 & \cdots & 0 \end{bmatrix}$$
$$\vdots$$
$$\mathbf{b}_{D-1} = \begin{bmatrix} 0 & 0 & 0 & \cdots & 1 \end{bmatrix}$$

The event, x, is formed from a linear combination of these coefficients.

$$\mathbf{x} = a_0\mathbf{b}_0 + \ldots a_{D-1}\mathbf{b}_{D-1}$$

However, alternative sets of basis functions exist which provide support for much of the energy in the original signal, but using far fewer than the original, D, coefficients. If, for example, a time signal trace is composed of samples of a pure sine-wave, and this is projected onto a set of basis functions consisting of samples of sine and cosine waves (carrying out a discrete Fourier transform), then a single coefficient could be used to retain all of the energy in the signal. Significant interest has been directed towards a group of basis functions collectively known as wavelets. The wavelet basis functions provide support for the parsimonious representation of a wide class of typical signals.

The wavelet transform is based on group theory and square integrable functions. It is an example of a time-scale decomposition where the scale decomposition is obtained by dilating the chosen analysing function. It guarantees locality since instead of breaking the signal down into global, harmonic functions, it is projected onto a space spanned by a series of local basis functions known as wavelets. These are each centred at a different position in the time axis, and decay to zero at a distance from their centre. In addition to the position, the scale of the wavelet may be altered. A particular local feature within a signal may be identified from the scale and position of the wavelets into which it is decomposed. Central to the theory of wavelets is the understanding that the wavelets represent a basis across complex space so that for any chosen wavelet function, there is only one possible decomposition for the signal being analysed [1]. The decomposition of a signal $x(t)$ into scaled wavelet coefficients $W_\psi^x(a,b)$ can be expressed as

$$W_\psi^x(a,b) = \int_{-\infty}^{\infty} x(t)\psi_{a,b}^*(t)dt \quad (2.0)$$

In this case $\psi_{a,b}^*(t)$ are the scaled wavelets, with dilation-a and translation-b, and $\psi^*(t)$ is the complex conjugate of $\psi(t)$. The translated wavelet is formed according the transformation in equation (2.1).

$$\psi_{a,b}^*(t) = \frac{1}{\sqrt{a}} \psi\left(\frac{t-b}{a}\right) \quad (2.1)$$

In order that the function $\psi(t)$ can be held to be a wavelet, it must satisfy the admissibility condition (2.2)

$$0 < C_\psi = \int_{-\infty}^{\infty} \frac{|\hat{\psi}(f)|^2}{|f|} df < \infty \quad (2.2)$$

Since it is desired that the transformed coefficients should demonstrate time domain locality, then the mother wavelet, $\psi(t)$, must also decay to zero at infinity (and hopefully much more rapidly), that is, it should have finite energy.

$$\int_{-\infty}^{\infty} |\psi(t)|dt < \infty \quad (2.3)$$

For compression purposes, the signal $x(t)$, should be able to be described accurately by a linear combination of as few terms as possible in the rotated space. Unfortunately little guidance is given in the literature to date on the selection of $\psi(t)$. However, a number of functions have been tested by researchers and shown to be suitable for decomposing sharp impact-type events. We use several of the most common: Daubechies 4-coefficient transform, Daubechies-8, Daubechies-12, Daubechies-16, Belykin-1, Coifman-2, Coifman-5, Symmlet-5, Symmlet-10 and Vaidymathan wavelets. These have all been described extensively in the literature [3,4]. A selection of orthogonal mother wavelets are presented in Figure 2.1 below.

These linear transforms are all rotations of the original data set, and projections onto an alternate set of axis. As such they may be represented by a matrix. They are energy preserving, so that all of the information contained in the original data is preserved in its transform. The transforms are reversible. The motivation for transformation of the data onto an alternate set of basis is that the transformed data matrix can be relatively sparse, with a suitably chosen set of wavelet coefficients. Much of the information contained in a vector can be represented in a few coefficients in this rotated space. This facilitates the feature extraction process.

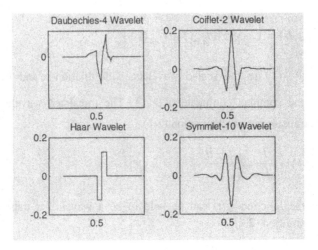

Figure 2.1- Orthogonal Wavelets

3. Feature Extraction Algorithms

The second stage in the construction of a classifier is the selection of a subset of features to use in the estimation of the class conditional density function. The task is to choose a small subset of features from amongst a relatively large group which describe the original signal. This subset of features can be formed by combining groups of the original features, by performing linear or non-linear operations on them, or by simply selecting a number of suitable features to form the subset. The "extracted" features should be relevant to the goal of assigning the waveforms to their correct classes, that is, those features which permit the best modelling of the class conditional probability density function over the training data set.

A linear operation has already been performed on the original time domain data by using the wavelet transform to project the data from the axis of the original attribute space onto the orthogonal axis of the Wavelet Transform domain, thereby rendering its representation in a more sparse form. The task in feature extraction is reduced to one of merely feature selection. This feature selection operation is concerned with choosing a small number of the elements from the transformed data matrix, $\hat{X}$, which will maximise the similarity of reduced feature vectors, $\hat{x}$, corresponding to waveforms of the same class. It must simultaneously minimise the similarity of those in different classes. The measure of this similarity or dissimilarity is related to the Euclidean distance between vectors. However, it depends in a very complex manner on the distribution of clusters of feature vectors in the reduced attribute space. The exact form of the dissimilarity measure will then, depend on the form of mapping used between the reduced feature attribute space and the classification target vectors. The most reliable method of assessing the suitability of a set of features is to carry out the classification process using the desired form of mapping, trained on the reduced data set.

To find an optimal set of d features from a candidate set of m features would require an extensive search of all possibilities. If we allow the same feature to be chosen many times, there are $\dfrac{m^d}{d!}$ possible ways of choosing the d features. The RBF classifier described in section 4 does not assign optimal weights to the contribution of individual features to the conditioned prediction of class membership. Deriving these weights would destroy the linear nature of the RBF training process. By allowing the same feature to be chosen several times, a mechanism is provided to weight heavily features which are significant to the classification decision.

If, as may typically be the case in many vibration or Acoustic Emission type applications, the original fully dimensioned data vector contains more than 1000 elements to represent an event, and we seek say 50 of these elements to use in the reduced feature vectors, then it is necessary to enumerate some 3.3×10^{85} possible subset combinations. Experience has shown that to train and test a 15 neuron RBF network of the type described in section 4 written in highly optimised C code, typically takes in the order of 2-3 seconds for a data set of some 100 training and test examples, using a Pentium 100MHz PC, 256KB on chip cache, 32Mbytes of RAM and running Windows 95™. An extensive search would then take 2×10^{78} years to complete.

3.1 Forward Selection

In order to avoid an extensive search, heuristics must be employed. Heuristics such as forward and backward selection will not necessarily lead to an optimal solution however. This is because the classification error rate does not necessarily decrease monotonically as features which are best are added progressively. Rather, the interactions of many subsets features would have to be considered. Floating search methods [5] yield better results but the computational effort required increases exponentially.

The current conclusions of the research community are summarised by Pudil *et al.* [7],

> "*Research and experience with Feature Selection has lead us to the conclusion that there exists no unique optimal approach to the problem*"

The simple forward selection algorithm employed is summarised below in psuedo code (Figure 3.1). A *fitness* function must be defined to assess the success of the chosen set of coefficients. In this case a very simple fitness function is used. The routine **Test_network** trains and tests a RBF network of the type described in section 4. An second data set is used to test the network distinct from that employed in the training process. **Test_network** returns a score representing the percentage of correct classifications made by the network. Given a classifier, $f(\mathbf{x})$ which maps a feature vector, $\mathbf{x}$, onto a one-of-N encoded classification, $\mathbf{t}$, with weight matrix, $\mathbf{W}$, a function, g, choosing D from M features, and a set of test data, $\mathbf{X}$, the fitness function has the form:

$$fitness(\Theta; g_\Theta : \Re^M \to \Re^D, f(\mathbf{x}; \mathbf{x} \in \Re^M, \mathbf{W}) \to \Re^N, \mathbf{X} \in \Re^{P \times M}) =$$

$$\frac{\sum_1^P \arg\{\max_j (g_\Theta(f_j(\mathbf{x}_i;)))\} = t_P}{P},$$

where $t_M = \mathrm{argmax}(\mathbf{t}_i)$.

Forward selection will produce sub-optimal results since it does not take into account the effect of grouping of features, but rather seeks to optimise the choice at each stage of the selection. Forward selection requires the enumeration of 5×10^4 possibilities, taking about 42 hours to pick the 50 most suitable elements from 1000 possibilities as above. Floating search methods permit back-tracking at each stage of the selection. As the amount of back-tracking is increased, floating search approximates to extensive search. Floating search will undoubtedly give better results in many cases, but is too time-consuming in this application.

```
d;                      -- features to add
X;                      -- original (p x m) training matrix
X_reduced=[];  -- reduced (p x d) training matrix
X2;                     -- original (p x m) test matrix
X2_reduced=[];          -- reduced (p x d) test matrix
for i=1:d
    bestscore=0.0
    for j=1:length_of_original_data_vector

        Train_network([X_reduced X[j] ])
        score=Test_network( [X2 X2[j]])

        if (score> bestscore)
            bestscore=score;
            candidates[i]=j;
        endif
    end
    X_reduced[j,:]=X[candidates[j],:]
    X_reduced2[j,:]=X2[candidates[j],:]
end
```

Figure 3.1 - Psuedo-code for Forward Selection of features

3.2 Genetic Algorithms

Genetic Algorithms (GAs) are a form of directed search which parallel the mechanism of Darwinian natural selection. They were developed by Goldberg in the 1970s and have attracted considerable attention in the engineering community over several years for the optimisation of a large number of problems such as optimisation of design parameters. Unlike many non-linear optimisation algorithms GAs do not require a mathematical model of the system that is to be optimised. Specifically, they make no use of the Hessian matrix, essential to any form of gradient descent, but which is all too often very difficult to obtain. Rather, they need only some function to assess the *fitness* of a particular solution. We use the same **Test_network** routine to arrive at a fitness as was used in the forward selection algorithm.

In a GA possible solutions are encoded as a string known as a *chromosome*, eg. 10111001. Each symbol in the string, which may be binary or a higher base, is known as a gene. An encoding strategy must be developed in order to describe the variables of a particular optimisation problem with such a string. Since the wavelet transform described in section 2 has 1024 coefficients, each coefficient may be represented as a 10-bit binary string. A set of fifty or coefficients can be written as a string of 500 zeros or ones. Having arrived at a representation the next stage is to generate a random initial population of possible solutions or genes. The number of genes in this population will vary according to the dimensionality of the space that is to be searched. The fittest genes must now be chosen from amongst the total population of encoded solutions. This is achieved by running the fitness function on each of the genes. The fitter genes are then used in a mating process to create the next generation of genes which will hopefully provide better solutions to the problem by retaining many of the most desirable characteristics from the earlier generation. The probability of a gene being chosen for breeding is equal to a measure of its fitness normalised over the entire population. When sufficient genes have been selected for mating they are paired up at random and their genes combined to produce two new genes. A position along the genes is chosen at random and the bits from each gene are switched.

Two mechanisms are used to prevent the population from stagnating. Perturbations are introduced into the population by adding some entirely new random genes. These are known as *new blood*. Genes may also be mutated by randomly inverting some of their bits with a small probability. It is usually not possible to be aware in advance of the best achievable fitness. The iterative process of breeding new generations is usually allowed to continue until the population is dominated by a few relatively fit genes which provide solutions which are good enough for purpose.

Many authors have investigated the use of genetic algorithms for feature selection. However, it has been noted that the lack of structure in their search strategies can lead to poor results in high dimensions.

4. Modelling $P(\mathbf{x}|C_j)$ with a RBF Network

A neural network model may be used to model the class conditional probability distribution. A Radial Basis Function network may be regarded as a form of linear non-parametric model. A general linear model of a system takes the form shown in equation 4.0 below. This gives an N-dimensional response $\mathbf{y}$ to an input vector $\mathbf{x}$. Each output in the vector $\mathbf{y}$ is a linear combination of the M specified basis functions $h_i(\bullet)$.

$$y_j(\mathbf{x}) = \sum_{i=1}^{M} w_{ij} h_i(\mathbf{x}) \quad (4.0)$$

In matrix notation this is represented as

$$\mathbf{Y} = \mathbf{HW} \qquad (4.1)$$

In a radial basis function network a number of functions are commonly used for $h_i(\bullet)$. However, the most common is the Gaussian. The basis functions then have the form

$$h_i(\mathbf{x}) = e^{\frac{-\|\mathbf{x}-\mathbf{c}_i\|}{\varphi^2}} \qquad (4.2)$$

To estimate the parameters in any form of model, a statistically over-determined set of output measurement data must be obtained. An experimental set-up is arranged and a number of recordings of input-output data sets made, $\left\{ (\mathbf{x}_i, \mathbf{y}_i) \right\}_{i=1}^{p}$. The number of data sets required, P, depends on the dimensionality of the input and output vectors and the order of the model. This determines the number of free parameters in the model. For larger models with more free parameters, a greater number of training examples must be used. In order to derive statistically significant weight values, the training data must be over-determined.

There are two main stages in the training of a Radial Basis Function network, namely:

- selection of centre locations for each basis function, $\mathbf{c}_i$;

- setting of hidden layer weights, w_{ij}.

Centre selection is undertaken in this case using the k-means clustering algorithm. According to this unsupervised clustering algorithm the initial centre locations are chosen randomly from the data points. An iterative approach is then applied whereby centre positions are nudged so that they are more closely aligned with the location of the centre of gravity of a cluster.

1. Loop over each input vector $\mathbf{x}_k$ from the training set.

2. Choose the nearest cluster centre, $\mathbf{c}_i$, to $\mathbf{x}_k$ and adjust centre position according to the equation:

$$\mathbf{c}_i(n+1) = \mathbf{c}_i(n) + \eta * (\mathbf{x}_k - \mathbf{c}_i) \qquad (4.3)$$

3. Loop over all input vectors, reducing η until changes in $\mathbf{c}_i$ are sufficiently small.

This algorithm is by no means optimal and is heavily dependent upon the initial choice of centre locations. However, results are much superior to those obtained by a purely random selection of centres.

Having chosen a set of basis functions $\{H_i : 1 \leq i \leq M\}$ to construct the model, the model is then fitted to the measured data by choosing appropriate values for w_{ij} (see equation 4.0). It is usual to define some form of "merit" or "cost" function for the model with a particular set of weight coefficient estimates, W, based on the square distance of the model predictions from the measured output values. Such a cost function is shown in equation (4.3) for a model with N outputs.

$$\aleph^2 = \sum_{j=1}^{N} \sum_{i=1}^{P} \left[y_{j_i} - \sum_{k=1}^{M} w_{kj} h_k(\mathbf{x}_i) \right]^2 \qquad (4.4)$$

Translating this into matrix notation we arrive at equation (4.5) for this cost function.

$$C(W) = \frac{1}{2}(Y - HW)^T (Y - HW) \quad (4.5)$$

The optimum selection for the weights in W occurs when all elements of the matrix C(W) are minimised.

$$\frac{\partial C}{\partial W} = H^T HW - H^T Y \qquad (4.6)$$

This occurs when the first differential of C(W) with respect to the weight matrix W is the zero matrix.

$$H^T HW = H^T Y \qquad (4.7)$$

Solved for W this gives,

$$\hat{W} = (H^T H)^{-1} H^T Y \qquad (4.8)$$

$(H^T H)^{-1}$ is normally evaluated using Gauss-Jordan elimination. However, in many applications the solution to this, the normal equation, is very close to singular. A zero pivot element may be encountered during the Gauss-Jordan elimination procedure giving rise to fitted parameters where the parameters are very large and cancel out almost precisely when the fitted model is evaluated at the training data points. However, they are unstable and valid predictions cannot be made away from these points. To avoid this numerical instability the matrix inversion may be performed using the Singular Value Decomposition, or a regularization term may be added to the cost function. In regularization the cost function of the network (see equation 4.4) is altered by addition of a penalty term $\Omega(\mathbf{y})$.

$$C(W) = \frac{1}{2}(Y - HW)^T (Y - HW) + \lambda \Omega(Y) \quad (4.9)$$

The parameter λ tempers the influence of the penalty term. The penalty term controls the variance of the network and sets a balance between an *a priori* form for the model with a strong bias towards prior knowledge of its form, suggested by relatively large values of λ, and a *a posteriori* determination of the model shape based primarily on the training data. In zeroth-order regularization, or weight decay as it is commonly referred to in the neural network literature, the regularization term has the form

$$\Omega(\mathbf{Y}) = \lambda \mathbf{W}^T \mathbf{W} \qquad (4.10)$$

This term is a measure of the smoothness or stability of the desired solution. It penalises large weight values which encourages solutions with small weight values. The additional smoothness is achieved since it is more likely that large weights will be required to produce an extremely rough output function. The effect of such a ridge regression technique is to reduce the effective number of degrees of freedom in the model, making it less flexible, so that it cannot fit so many functions. The advantage is that the model is more stable. If we were to minimise C(W) with respect to this term alone then we would arrive at a solution that was very smooth, but that had nothing to do with the shape of the measured data. Other regularization terms may be used rather than (4.10) depending on *a priori* knowledge of the shape of the function [Press]. However, in this application zeroth-order regularization is used with λ set to a value of 0.1.

The model described will not in fact produce Bayesian probabilities, rather, likelihood ratios for class membership. Since the model outputs are only used to pick the class with the maximum likelihood, it is not important that the outputs cannot strictly be regarded as Bayesian probabilities.

5. Experimental Validation

Cracking of pencil leads has been suggested as providing a useful simulation of the type of Acoustic Emission associated with the failure of composite materials. It is well known that such Acoustic Emission signals are extremely difficult to classify. A range of pencil leads were snapped by pressing against the base plate of an Acoustic Emission transducer and the Acoustic Emission waveforms recorded. Six types of pencil were used. These are detailed in Table 1.

Class	Description of type of pencil lead
1	Sharpened 2B graphite pencil, diameter 1mm
2	Pop-up HB graphite, diameter 0.5mm
3	Pop-up 2H graphite, diameter 0.5mm
4	Pop-up 4H graphite, diameter 0.5mm
5	Pop-up HB graphite, diameter 0.7mm
6	Sharpened HB graphite pencil, diameter 1mm

Table 1 - Types of pencil lead tested

A transient data capture board with suitable anti-aliasing filters was used to record data from a Stresswave™ Acoustic Emission transducer modified to remove the enveloping circuitry on its output to preserve frequency content information. Data was sampled at 780KHz then subsampled at 390KHz. Figure 5.1 shows an example of a pencil cracking event from class 5.

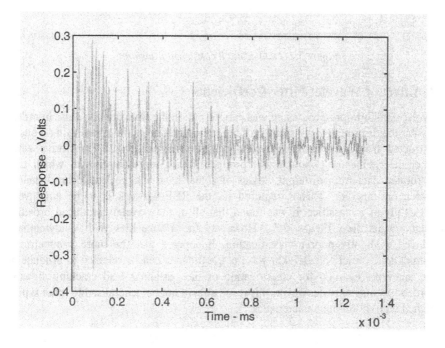

Figure 5.1 - Example of class 5 waveform

The training data was divided into two groups. One group of data was used to set the free parameters of the network, and the other group to calculate a percentage success rate in classification. A wavelet transform using the wavelet filter coefficients described in section 2. Figure 5.2 shows an example of a Daubechies 8-coefficient wavelet transform of the pencil snapping data shown previously.

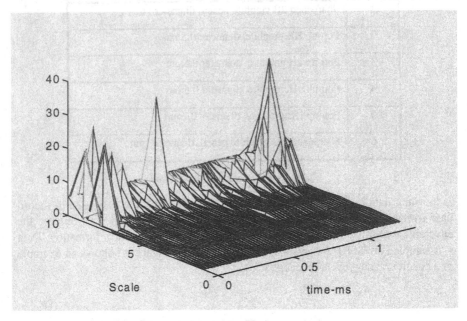

Figure 5.2- Daubechies 8 coefficient transform

5.1 Effect of Wavelet Filter Coefficients

Wavelet transform pre-processing was carried out on the time domain data, in order to assess the discriminatory potential of the representation of the signal using a basis constructed from the popular wavelets described in section 2. Transforms with greater discriminatory potential will produce a data representation in which the transformed data from different classes of signal have a large Euclidean distance between one another. This is required if the RBF network is to be employed successfully as a classifier. It was found that all of the wavelets produced broadly similar results (see Figure 5.3). However, the Daubechies and Vaidymathan produced a slightly poorer classification. It appears that the more symmetrical Symmlet and Coiflet and Belykin wavelet coefficients can be selected to produce a more apt representation for classification of the particular lead cracking signals recorded for this paper. However, this result should not be generalised to other types of signal with different characteristics.

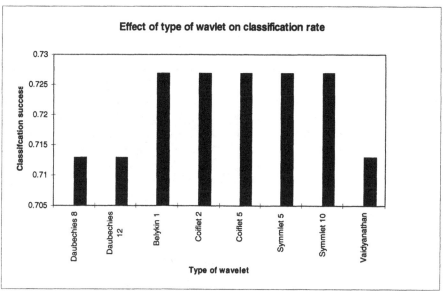

Figure 5.3 - Effect of wavelet filter on classifcation success

5.2 GAs or Forward Selection of Features

Figure 5.4 compares the accuracy of a classifier based on wavelet coefficients chosen using a forward selection (FS) algorithm or, alternatively, with a genetic algorithm. A population size of 100 genes was used, each gene having a length of 300 bits (30 features, where each feature is chosen from 1024 possibilities). This ensures that the runtime for the GA is approximately the same as for the Forward Selection procedure.

It is intuitive that groups of coefficients will together form very powerful indicators of class. This suggests that forward selection of features will be extremely sub-optimal. However, in all cases the FS outperformed the GA. Evidently the topology of the space to be searched by the GA is too flat or possess too few clearly defined structures to allow for effective searching with a GA. This may be governed by the lack of discernible pattern in the coefficients of the transformed data that result in a discriminatory representation.

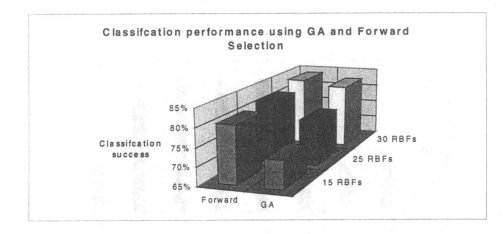

Figure 5.4 - GAs and Forward Selection for feature selection

When a very complex classifier is employed, with many RBF neurons, the performance of the GA approaches that of FS. With this form of overly complex model, the choice of features is no longer critical. If it is important that a parsimonious model of the attribute space to classifcation vector mapping is to be produced, for use for example in demanding real-time signal identification applications, then it may be better to use the FS to choose the model inputs. However, if sufficient computer power is available to use a large RBF network, then GAs can deliver a trained classifier in a fraction of the time taken by the rigorous FS search.

A classification accuracy of 83% may not appear to be very great. However, the data set was chosen specifically because it has proven extremely difficult to classify. The classification success rates with the use of the wavelet transform and forward selection should be contrasted against taking the first 50 principal components of the data, which gave a best classification rate of 32%. Classification using the whole 1024 elements of time domain data was time-consuming and achieved a success rate of 38%.

6. Conclusions

A description has been given of a classification system for high frequency waveforms. The analysis of such waveforms is a common task in fault identification. The classifier uses wavelet transforms for signal pre-processing to produce a representation which may more readily be compressed. A comparison is then presented of the use of a forward selection algorithm and a GA for feature picking. A Radial Basis Function network was employed to model the *posteriori* probability density function. The classifier was then applied to the identification of pencil leads snapping Acoustic Emission data. The Forward Selection based classifier achieved a better success rate in the classification of the Acoustic Emission pencil snapping data than a model with inputs chosen using a Genetic Algorithm or a classifier working on the raw time domain data.

Research is required in order to understand better the nature of the relationship between wavelet coefficients which results in their combined use as classifier inputs. It may then be possible to derive GA fitness functions which explicitly optimises the selection of groups of features.

Acknowledgements

The authors would like to thank Dr. Keith Worden, University of Sheffield, both for the code for his SGA genetic algorithm which formed the basis of the GA used in this paper and for his advice and comments. Funding has been provided by Rolls-Royce Applied Science Labs., Derby, and the EPSRC under the Engineering Doctorate Programme.

References

1. Bishop, C. Neural Networks for Pattern Recognition, Oxford University Press, 1995.

2. Broomhead, D.S., Lowe, D., "Multi-variable Functional Interpretation and Adaptive Networks", Complex Systems, Vol. 2, 1988, pp.321-355.

3. Chui, Ch., K., An Introduction to Wavelets. Wavelet Analysis and its Applications, Vol. 1, Academic Press, Boston, 1992.

4. Daubechies, I., "The Wavelet Transform, Time-frequency Localization and Signal Analysis", I.E.E.E. Trans. Information Theory, Vol . 36, 1990, pp.961-1005.

5. Kittler, J., "Feature Selection and Extraction", Handbook of Pattern Recognition and Image Processing, Eds.Young, T.Y., Fu, K.S., Academic Press, 1986, pp. 60-81.

6. Newland, D.E., *An Introduction to Random Vibrations, Spectral and Wavelet Analysis*, 3rd Ed., Longman Scientific and Technical, Essex, 1994.

7. Pudil, P., Novovicová, J., Ferri, F., "New Tools for Knowledge Guided Approach to Statistical Pattern Recognition", *AISB96 Workshop*, 1 April, 1996, Brighton, UK.

8. Press, W.H., *et al.*, Numerical Recipes in C, Cambridge University Press, 1994, pp. 808-812.

The Contruction and Evaluation of Decision Trees: A Comparison of Evolutionary and Concept Learning Methods

H.C. Kennedy, C. Chinniah, P. Bradbeer†, L. Morss*

Department of Computer Studies,
Napier University, Edinburgh, Scotland EH14 1DJ

†email pvgb@dcs.napier.ac.uk   * email les@dcs.napier.ac.uk

Abstract The CALTROP program which is presented in this paper provides a test of the feasibility of representing a decision tree as a linear chromosome and applying a genetic algorithm to the optimisation of the decision tree with respect to the classification of test sets of example data. The unit of the genetic alphabet (the "caltrop") is a 3-integer string corresponding to a subtree of the decision tree. The program offers a user a choice of mating strategies and mutation rates. Test runs with different data sets show that the decision trees produced by the CALTROP program usually compare favourably with those produced by the popular automatic induction algorithm, ID3.

1. Introduction

Decision trees are widely used to represent schemes for classifying entities according to values of attributes. The decision trees themselves have wide application, for instance in medical and other diagnostic systems (see for example, Muggleton [9]). A great deal of attention has been given to algorithms for induction which, given a training set of examples each of which is described by the values of a number of attributes and an outcome, will automatically build decision trees that will correctly classify not only all the examples in the training set, but unknown examples from the wider universe of examples of which the training set is presumed to provide a representative sample. Automatic induction systems have recently been reviewed by Langley [5].

As an example, consider the data in Table 1, in which a set of students is described in terms of the values of attributes which represent items of their outer clothing and appearance. The requirement is to build a decision tree which will classify each student according to gender using the values of these attributes. Table 1 shows six exemplars of students with different combinations of attribute values. Note that there are only two outcomes - *male* and *female* - and that there is a many-to-one relationship between attribute-value combinations and each outcome. Note also that there is no one attribute that will separate the exemplars into two subsets each

	Attribute 1	Attribute 2	Attribute 3	OUTCOME
Exemplar of Student	Wears a tie? (Y/N)	Long hair? (Y/N)	Lower Garment: Skirt-like (S) or Trousers (T)	Gender (M/F)
Student 1	Y	N	T	M
Student 2	N	Y	S	F
Student 3	Y	Y	T	M
Student 4	N	Y	T	F
Student 5	N	N	S	F
Student 6 (a kilted Scot!)	Y	Y	S	M

Table 1: Examples of students to be classified into gender according to attributes representing outer appearance.

of which only contains students of the same gender. Many decision trees can be constructed which would correctly classify all of the exemplars in Table 1: fig. 1 gives two examples. Tree (a) is appropriate if the majority of students follow exemplars 1 and 2. However, if the class contains many Scotsmen in kilts, and a large number of female students wear jeans, then some other attribute should head the tree (Tree b). The optimum tree for a given example set depends critically on the frequency distribution of the exemplars.

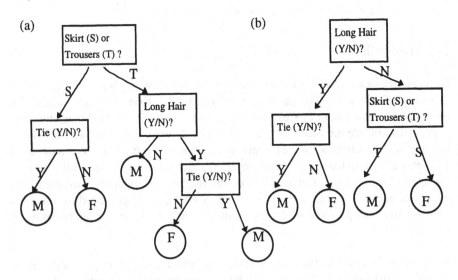

Fig. 1   Example of a simple decision tree

This set of exemplars demonstrates important aspects of decision tree construction:

- There must be no "clashes" between exemplars: induction procedures cannot deal with examples which have identical attribute values, but different outcomes.

- Many different decision trees can be constructed which would correctly classify all the examples in the example set. The search space of candidate decision trees which must be examined in order to identify the optimum tree for classification of a given data set grows very quickly with the number of attributes which contribute to the identification. Langley [5] states that where there are a attributes with v values each, the number of possible correct trees grows as

$$\prod_{i=0}^{a=1}(a-i+1)^{v^i} \qquad \text{(Langley [5]).}$$

Genetic algorithms are effective mechanisms for exploring large search spaces (Goldberg [6]) and in this paper we examine the feasibility of using such an algorithm to detect an optimum decision tree in the set of trees that will correctly classify the data in a simple example set. For simplicity, all attributes are Boolean and the outcomes of classification are also binary (*True* or *False*). We assume that the optimum decision tree is the one which will classify the example set, and by extension the universe of entities of which the example set is a representative sample, using the least number of decisions. Candidate decision trees are compared by the number of decisions required to classify correctly all the examples in the set. Trees which do not correctly classify all the examples are heavily penalized.

Automatic induction programs which build decision trees generally utilize machine learning algorithms such as the well-known ID3 (Quinlan, [10]) and its more recent modifications such as C4.5 (Quinlan, [11]). ID3 constructs decision trees from the top down, selecting the first attribute on the basis of an information metric, and proceeding recursively down the branches of the nascent tree until all examples are correctly classified. Decision tree induction methods have been reviewed by Mingers [8] and more recently by Buntine and Niblett [1]. We have used the ID3 algorithm as a basis of comparison for the performance of the genetic algorithm. For a good description of the ID3 algorithm, see Ginsberg [4], pp313-317.

2. Representation of a Decision Tree as a Linear Chromosome

The principal problem in representation is to map the two-dimensional decision tree onto a linear genetic chromosome. The approach adopted was to consider a binary decision tree as being composed of a number of unit subtrees, each consisting of a root node and two branches. We chose to refer to these subtrees as "caltrops", from their superficial similarity to the spiked devices used in mediæval warfare to disable charging horses. The root node of a caltrop represents one of the attributes by which

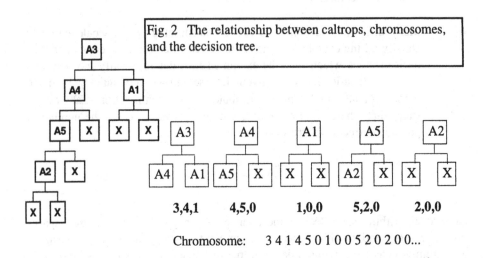

Fig. 2   The relationship between caltrops, chromosomes, and the decision tree.

the examples are classified. The two daughter nodes can likewise be attributes, but additionally can represent leaf nodes, or terminals, to allow for those cases where in the decision tree the subset at the node is fully classified. In a caltrop, the integer 0 corresponds to a leaf node, and other integers refer to attributes. The integer at the root node of a caltrop is not allowed to take the value 0 since the root of a subtree cannot itself be a leaf.   Each gene on the chromosome is a string of three integers which maps onto a caltrop. Fig. 2 shows how a complete decision tree can be represented in a chromosome.

A chromosome is made up of a set of "caltrops". Each caltrop is an indivisible unit in this representation and cannot be split by crossover. The chromosomes are linear, but to make their structure more apparent we represent them in this paper as tables in which each row represents a single caltrop (e.g. fig. 4).

2.1 Mapping a Chromosome onto a Decision Tree

Each population was initialised by randomly generating a population of (normally 40) chromosomes. A chromosome of n caltrops consists of a string of $3n$ integers each in the range 0 - q where q is the number of attributes. The integers corresponding to the head nodes are constrained to the range 1 - q since a head node cannot be terminal. Subsequent populations contained chromosomes modified by the operations of crossover and mutation.

A chromosome was interpreted into a decision tree according to the following procedure:

- Take the first row (caltrop) of the chromosome as the root of the decision tree: the first integer is the attribute to be used as the root node, the second integer is the attribute to be used as the left branch and the final integer in the caltrop is used as the attribute at the right branch.

- Search the chromosome for the **first** caltrop whose head is the left daughter node of the root node.

- Continue to build the tree in a depth-wise manner until a terminating condition is found. Then backtrack, building the tree from right-hand daughter nodes.

- Backtrack till a right-hand daughter node is found; then continue to build the tree as in Step 3. Iterate steps 3 and 4 until all daughter nodes have been built from.

A terminating condition is found when one of the following conditions becomes true:

- a leaf node is found, represented by a zero in the caltrop.

- all attributes have been used in the path. The node is branched into two leaves - since all attributes have been used, if these leaves do not resolve the outcomes then the example set itself cannot be classified using the given attributes.

2.2 Example of the Interpretation of a Chromosome into a Decision Tree

Consider the chromosome represented by the integer string which begins with the sequence:

$$3\ 4\ 1\ 4\ 5\ 1\ 1\ 0\ 0\ 5\ 2\ 0\ 2\ 0\ 0\ ....$$

| | Caltrop: | #1 | #2 | #3 | #4 | #5 |

The chromosome is visualised as a set of "caltrops" each represented by a string of three integers:

#1	3	4	1
#2	4	5	1
#3	1	0	0
#4	2	0	0
#5	5	2	0
#6	...	...	...

Fig. 3 shows how the set of caltrops that makes up the chromosome is interpreted into a decision tree. Each box in the figure delineates one of the five caltrops represented in the chromosome, and the sequence in which they are used to construct the tree is indicated by the numbers in the left or right upper corner. The process is depth-first, proceeding down a left branch until a terminal node is encountered. Backtracking then takes place to assign caltrops to right-hand daughter nodes. The process is explained in more detail below:

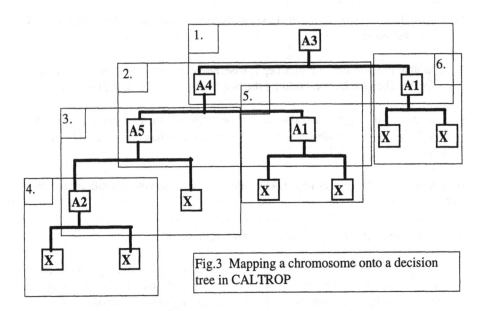

Fig.3 Mapping a chromosome onto a decision tree in CALTROP

1. The first caltrop in the chromosome is taken to be the head of the decision tree.

2. The tree is built up starting with the left-hand daughter of the root node. In this case the attribute at the left daughter node is A4. The chromosome is searched for the first caltrop whose root is A4. In this case it is caltrop #2 in the chromosome.

3. 4. The decision tree continues to be built up in a depth-first manner working down the left branch of each node. As each caltrop is added to the tree, the chromosome is searched for the first caltrop whose root matches the left hand daughter of the node above.

4. The daughter nodes of caltrop #4 both contain 0, which is interpreted as a leaf node (represented by X in the decision tree). Branch construction can therefore proceed no further in this direction. The treebuilder backtracks back up the branch until it comes to a node whose right-hand daughter is not a leaf node X.

5.

The first such node is A4, which was put in place by step 2 of the tree construction process. As in other steps, the chromosome is searched for the first caltrop whose head matches the daughter node (in this case A1). The caltrop found is #3. The daughter nodes of this caltrop are both leaf nodes, so that at this point the left subtree of the root node of the decision tree has been completed. Backtracking now returns to the root node to build the right subtree, beginning with A1.

6.

The right-hand daughter of the root node is A1, so the chromosome is searched for the first caltrop which is headed by A1. Once again the caltrop found is #3; however, there is no restriction on using the same caltrop more than once in a decision tree, provided that an attribute is not duplicated in a single path from the root to a leaf. Since the daughter nodes of caltrop #3 are both leaf nodes, no further tree construction can take place - the decision tree represented by the chromosome is complete.

2.3 Special Cases in the Interpretation of the Chromosome

Two types of terminating condition do not correspond to satisfactorily completed trees and a design decision was taken to resolve them arbitrarily:

- the chromosome does not contain a caltrop whose root integer matches the current requirement.

- a repeated attribute has occurred in a decision path. The repeated occurrence is redundant.

In either case, the path was terminated with a pair of leaf nodes, one for each value of the terminal attribute. The leaves may contain mixtures of outcomes, and such a tree will not correctly classify examples. Trees which do not correctly classify the examples are heavily penalized in the fitness function.

2.4 Properties of the Decision Tree

Of the decision trees generated by the random initial population chromosomes, and those subsequently created by crossover and mutation, many will not classify the example set. Within this paper we refer to a decision tree that does not correctly classify the example set as **illegal**. A tree in which each leaf node holds only examples of the same outcome is referred to as **legal**. The **efficiency** of trees is compared using the total number of decisions (i.e. the total number of nodes traversed) required to classify all of the examples in the example set.

3. Operation of the CALTROP Program

Crossover is carried out **between** caltrops on the chromosome: the caltrops are regarded as units in the genetic alphabet (fig 4). The new chromosomes resulting from cross-over can be interpreted into decision trees, although these will not necessarily be legal (as previously defined).

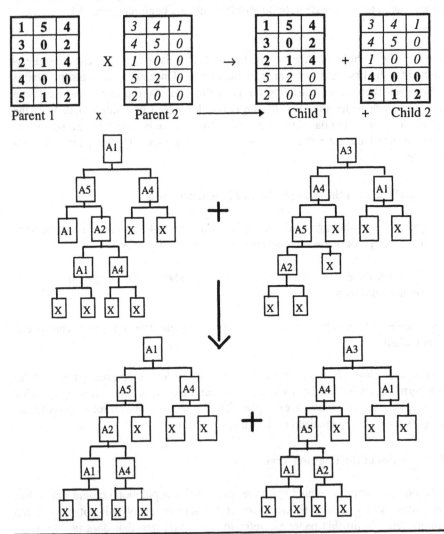

Fig. 4 Crossover implemented in the Caltrop representation. X represents a leaf node, A*N* an attribute.

Mutation is implemented by the replacement of a caltrop in a chromosome chosen at random by a randomly generated triplet. The user can select one a range of mutation rates from 1 caltrop in 250 to 1 in 2500.

The selection of pairs for crossover is implemented by a variety of solution methods from which the user can choose at run-time. Currently implemented are *roulette wheel* (Goldberg [6]), *tournament selection*, described for example in Goldberg [7]), and *assortative mating* - a variation on genetic invariance (Culberson, [3]). Elitism can also be selected. Assortative mating with elitism was employed for the experiments described in the Appendix.

In the current implementation, the user is offered populations of up to a maximum of 40 chromosomes, each a string of 72 integers, or 24 caltrops. This maximal population was used in each test run.

4. Testing and Results

4.1 Evaluation of Decision Trees

A simple metric was employed to evaluate the efficiency of the decision trees. This was the number of decisions which would be required to classify all the members of the example set (Cheng *et al.*, [2]). With 40 members of the example set and 4 attributes, the fitness value of the least efficient legal decision tree is 4 x 40 = 160. After all the examples in the set were classified by a given decision tree, the number of outcomes at each leaf node was evaluated. Where a leaf node contained only a single outcome, the number in the node was multiplied by the number of decisions taken to reach that node. Where the leaf node contained both outcomes, the number of the less frequent outcome was multiplied by a penalty factor and added to the fitness function for the decision tree. Following the results of preliminary test runs, the penalty factor was set at a value of 100 for all of the experiments described in the Appendix.

4.2 Test Data

Test sets of data were constructed in such a way that the example sets could be contrived to give decision trees of markedly distinct form. A set of examplars was created to build data sets on which to test the performance of CALTROP; these are shown in Table 2. The exemplars are a set of ten integers: {-6,-5,-3,-2,2,5,6,10,12,15}. The subset {-6,-5,2,12} is arbitrarily associated with the outcome "1", and the non-intersecting subset {-3,-2,5,6,10,15} is arbitrarily associated with the outcome "0". The four attributes are simple arithmetical tests, and the values are Boolean results TRUE or FALSE. The Appendix shows five example sets made up of these exemplars represented at different frequencies. These example sets were chosen to give rise to decision trees with a variety of forms.

The four attributes were as follows:

- Is the integer greater than 0 ?
- Is the integer greater than 9 ?
- Is the integer exactly divisible by 2 ?
- Is the integer exactly divisible by 3?

Integer	>0	>9	/2	/3	Outcome
-6	F	F	T	T	1
-5	F	F	F	F	1
-3	F	F	F	T	0
-2	F	F	T	F	0
2	T	F	T	F	1
5	T	F	F	F	0
6	T	F	T	T	0
10	T	T	T	F	0
12	T	T	T	T	1
15	T	T	F	T	0

Table 2. Attributes and Outcomes of the Sample Data.

4.3 Results and Analysis

Table 3 shows the number of decisions required to classify each test set for the most efficient decision trees constructed using either ID3 or CALTROP.

Five data sets, each of 40 examples drawn from the exemplars, were made up with different frequencies of the 10 integers, so that each gave rise to a different decision tree. These data sets are given in the Appendix together with details of the tests carried out. For each data set, a decision tree was also constructed using ID3. The calculations of the ID3 algorithm were carried out with a spreadsheet.

For every example set, CALTROP found a legal decision tree that would correctly classify the examples. The number of generations taken to find a legal tree varied from 1 to greater than 500. In 3 out of the 5 test sets, CALTROP found a tree which was at least as efficient as that constructed using ID3. In these 3 cases, CALTROP found the most efficient tree which could be shown to classify the set. Assortative mating generally found a legal tree more quickly than tournament selection.

The results were considered to be indicative of the feasibility of the caltrop procedure. In each, the result with CALTROP is the best of a number of runs and it was felt inappropriate to apply statistical analysis.

Test Set	ID3	CALTROP
1	128	129
2	124	124
3	135	129
4	121	121
5	80	93

Table 3. Comparison of the efficiency of trees produced by ID3 and by CALTROP.

5. Conclusions

The CALTROP program demonstrates that legal and efficient decision trees can be built using a genetic algorithm. The decision trees built by this program are in many cases able to classify a test data set with fewer decisions than the decision tree built using the popular ID3 algorithm. The most efficient trees built by CALTROP often differed from those constructed using ID3, and the difference often extended to the attribute at the root node.

This observation reflects a difference in the basic approaches adopted by an automatic induction algorithm such as ID3 and the genetic algorithm. ID3 adopts a local top-down approach, selecting the attribute which is most efficient (by the criterion of information content) to partition the whole example set for the root node before proceeding to recursively partition the subsets. But the attribute which most effectively partitions the entire set is not necessarily the attribute which heads the most efficient decision tree. ID3 is unable to backtrack to redo previous choices. A well-constructed evolutionary approach, by contrast, adopts a "holistic view" of the decision tree: the entire tree is coded in the chromosome and subject to evolutionary improvement. This may explain why, in several of the test cases, the genetic algorithm was able to finally generate a more efficient tree than ID3.

The tendency of the CALTROP program to produce decision trees headed by Attribute 3 requires further investigation. This could represent premature convergence due to the choice of penalty function rather than evolution towards an optimum structure. The maintenance of diversity of trees in the population depends critically on the penalty applied to illegal trees: too high a penalty will have the effect of committing the population to the root attribute that most reduces the penalty applied to the trees generated. One possibility which might be investigated is to allow the penalty to increase over time. Other future work would be to test the CALTROP system against the standard data sets used for machine learning system.

A potential advantage of the genetic algorithm approach to decision tree construction lies in the possibility of dealing with examples from a distribution which is not constant, but changes in frequency over time. ID3 and other induction algorithms produce a reasonably well-optimised tree according to the relative frequencies of the examples in the test set. If however, these frequencies change in the "real-world", the decision tree will no longer be appropriate and it will be necessary to reconstruct the tree, or suffer the overhead incurred by the use of a non-optimal decision tree. The genetic algorithm approach, however, maintains a population of chromosomes whose decision trees are not identical and which will vary in the frequency distribution for which they are optimal. One approach to classification using genetic algorithms might be to maintain a population of distinct decision trees and select an optimum tree according to the current distribution of the example set. This will be a subject of future work.

This work was carried out in partial fulfilment of the requirements for the degree of Master of Science in the Department of Computer Science, Napier University, Edinburgh.

References

[1] Buntine W. and Niblett T. (1992) "A Further Comparison of Splitting Rules for Decision-tree Induction" *Machine Learning* **8** , pp 75-86

[2] Cheng J., Fayyad U.M., Irani K.B., and Quian Z. (1988) "Improved Decision Trees: A generalized version of ID3" *Proc. 5th International Conf. on Machine Learning,* University of Michigan

[3] Culberson J. (1992) *Genetic Invariance: a new paradigm for Genetic Algorithm design* University of Alberta Technical Report TR92-02
(http://web.cs.ualberta.ca/~joe/Abstracts/TR92-02.html)

[4] Ginsberg M. (1993) *Essentials of Artificial Intelligence* Morgan Kaufman, San Mateo

[5] Langley P. (1996) *Elements of Machine Learning* Morgan Kaufman, San Mateo

[6] Goldberg D.E. (1989) *Genetic Algorithms in Search, Optimisation, and Machine Learning* Addison Wesley, Wokingham, England

[7] Goldberg D.E. (1991) "A Comparative Analysis of Selection Schemes used in Genetic Algorithms" in Rawlins G. (ed.) *Foundations of Genetic Algorithms* pp. 69-93 Morgan Kaufman, San Mateo

[8] Mingers J. (1989) "An Empirical Comparison of Selection Measures for Decision Tree Induction" *Machine Learning* **3**, pp. 319-342

[9] Muggleton S. (1990) *Inductive Acquisition of Expert Knowledge* Addison Wesley, Wokingham, England

[10] Quinlan J.R. (1986) Induction of Decision Trees *Machine Learning* **1**, pp. 81-106

[11] Quinlan J.R. (1993) *C4.5: Programs for Machine Learning* Morgan Kaufman, San Mateo

Appendix

Each test set consists of 40 examples made up of different frequencies of the 10 integer exemplars used. For each example, the decision tree generated by ID3 (calculated using a spreadsheet) is shown above, and a typical tree produced by one run of CALTROP is given below. Since these tests were carried out to determine whether the CALTROP system was capable of producing more efficient trees than ID3, a variety of run conditions were used to generate the trees. Typically, mutation rates of 1 in 500 were employed, with assortative mating and elitism. A penalty value of 100 was applied to the fitness of any chromosome that produced an illegal tree.

For each test, the system was run until a legal decision tree was obtained whose efficiency did not improve in successive generations. A wide variety of generations was found to be required. Some runs achieved a decision tree after five generations which did not then change; in other runs the tree continued to improve after several thousand generations.

Test Example Set 1:

Integer	Occurr -ences	Outcom e
-6	5	1
-5	5	1
2	5	1
12	5	1
-3	3	0
-2	3	0
5	4	0
6	3	0
10	3	0
15	4	0

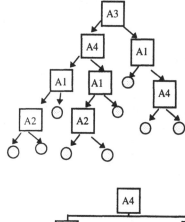

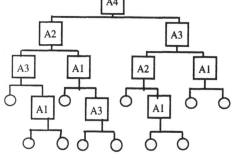

Test Example Set 2:

Integer	Occurr -ences	Outcome
-6	4	1
-5	4	1
2	3	1
12	3	1
-3	5	0
-2	4	0
5	4	0
6	4	0
10	4	0
15	5	0

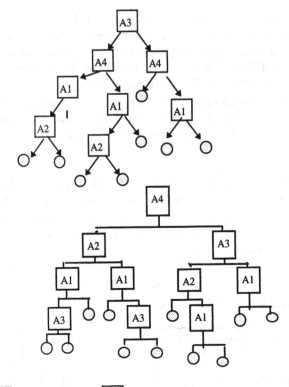

Test Example Set 3:

Integer	Occurr- ences	Outcome
-6	2	1
-5	2	1
2	7	1
12	8	1
-3	4	0
-2	4	0
5	3	0
6	3	0
10	3	0
15	3	0

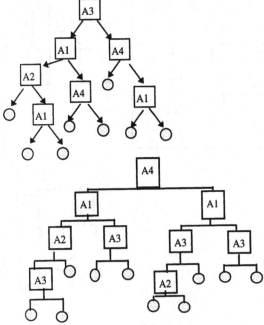

Test Example Set 4:

Integer	Occurr-ence	Outcome
-6	7	1
-5	3	1
2	4	1
12	2	1
-3	0	0
-2	4	0
5	3	0
6	3	0
10	3	0
15	7	0

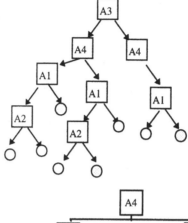

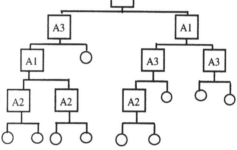

Test Example Set 5:

Integer	Occurr-ence	Outcome
10	19	0
6	6	0
5	3	0
2	11	1
-5	1	1

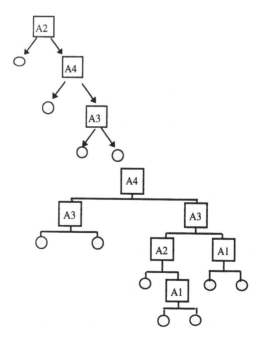

Parallel Distributed Genetic Programming Applied to the Evolution of Natural Language Recognisers

Riccardo Poli

School of Computer Science
The University of Birmingham
Birmingham B15 2TT, UK
E-mail: R.Poli@cs.bham.ac.uk

Abstract. This paper describes an application of Parallel Distributed Genetic Programming (PDGP) to the problem of inducing recognisers for natural language from positive and negative examples. PDGP is a new form of Genetic Programming (GP) which is suitable for the development of programs with a high degree of parallelism and an efficient and effective reuse of partial results. Programs are represented in PDGP as graphs with nodes representing functions and terminals, and links representing the flow of control and results. PDGP allows the exploration of a large space of possible programs including standard tree-like programs, logic networks, neural networks, finite state automata, Recursive Transition Networks (RTNs), etc. The paper describes the representations, the operators and the interpreters used in PDGP, and describes how these can be tailored to evolve RTN-based recognisers.

1 Introduction

In Genetic Programming [11, 12] programs are expressed as parse trees to be executed sequentially in the standard depth-first evaluation order. This form of GP has been applied successfully to a large number of difficult problems like automated design, pattern recognition, robot control, symbolic regression, music generation, image compression, image analysis, etc. [11, 12, 9, 10, 1, 18].

When appropriate terminals, functions and/or interpreters are defined, standard GP can go beyond the production of sequential tree-like programs. For example using cellular encoding GP can be used to develop (grow) structures, like neural nets [6, 7] or electronic circuits [15, 13], which can be thought of as performing some form of parallel analogue computation. Also, in conjunction with an interpreter implementing a parallel virtual machine, GP can be used to translate sequential programs into parallel ones [24] or to develop some kinds of parallel programs [3, 2, 23]. However, all these methods either are limited to very special kinds of parallelism or use indirect representations which require computationally expensive genotype-to-phenotype mappings.

This paper describes an application of Parallel Distributed Genetic Programming, a new form of GP which is suitable for the development of programs with a high degree of parallelism and distributedness. Programs are represented in PDGP as graphs with nodes representing functions and terminals, and links representing the flow of control and results. In the simplest form of PDGP, links are directed and unlabelled, in which case PDGP can be considered a generalisation of standard GP (trees are special kinds of graphs). However, PDGP can use more complex (direct) representations, which allow the development of symbolic, neuro-symbolic and neural networks, recursive transition networks, finite state automata, etc.

Like GP, PDGP allows the development of programs of any size and shape (within predefined limits). However, it also allows the user to control the degree of parallelism of the programs to be developed. In PDGP, programs are manipulated by special crossover and mutation operators which, like the ones used in GP, guarantee the syntactic correctness of the offspring. This leads to a very efficient search of the space of possible parallel distributed programs.

In previous work [19, 20, 17] we have studied some of the representation capabilities of PDGP and compared PDGP with standard GP on simple problems. In this paper we will show how the representations, operators and interpreters used in PDGP can be tailored to solve a much harder problem: the induction of a recogniser for natural language from positive and negative examples. A recogniser for natural language is a program that given a sentence (say in English) returns true if the sentence is grammatical, false otherwise. The problem of inducing recognisers (and parsers) from actual sentences of a language, also known as language acquisition, is a very hard machine learning problem (see [21, pages 443–451] for a survey on the topic).

A limited amount of work has been done on evolutionary algorithms applied to the problems of natural language processing. A very recent review of these is presented in [5]. To the author's knowledge only two applications of GP in this area have been published to date [22, 16], none dealing with the problem of natural language recognition. Somehow more closely related to this topic is the work described in [4, 17] on inducing, from positive and negative example, deterministic finite state automata capable of recognising simple regular languages.

Deterministic finite state automata are well suited to build recognisers for this kind of languages. However, they are not particularly suited to represent the natural recursivity of natural-language grammars. Indeed, the work in [4, 17] considered very simple non-recursive languages, like L=a*b*a*b* (consisting of all sentences with 0 or more a's followed by 0 or more b's, etc), which have nothing to do with the complexity of natural language. For this reason we have decided to use PDGP to evolve recognisers based on Recursive Transition Networks (RTNs). RTNs are extensions of finite state automata, in which the label associated to a link can represent either symbols of the language (like in standard finite state automata) or other RTNs (possibly including the RTN containing the link).

The paper is organised as follows. Firstly, in Section 2, we describe the representation, the genetic operators and the interpreter used in PDGP. Then, we illustrate how PDGP can be used to solve the problem of inducing recognisers for natural language from positive and negative examples (Section 3). Finally, we draw some conclusions in Section 4.

2 PDGP

2.1 Representation

Standard "tree-like" programs can often be represented as graphs with labelled nodes (the functions and terminals used in the program) and oriented links which prescribe which arguments to use for each node when it is next evaluated. Figure 1(a) shows an example of a program represented as a graph. The program implements the function max(x * y, 3 + x * y). Its execution should be imagined as a "wave of computations" starting from the terminals and propagating upwards like in a multi-layer perceptron.

This form of representation is in general more compact and efficient than a tree-like one and can be used to express a much bigger class of programs which are parallel in nature: we call them *parallel distributed programs.*

PDGP can evolve parallel distributed programs using a direct graph representation which allows the definition of efficient crossover operators always producing valid offspring. The representation is based on the idea of assigning each node in the graph to a physical location in a multi-dimensional grid with a pre-fixed shape.

This representation for parallel distributed programs is illustrated in Figure 1(b), where we assumed that the program has a single output at coordinates (0,0) (the y axis is pointing downwards) and that the grid includes $6 \times 6 + 1$ cells. We have also assumed that connections between nodes are upwards and are allowed only between nodes belonging to adjacent rows, like the connections in a standard multi-layer perceptron. This is not a limitation as, by adding the identity function to the function set, any parallel distributed program can be described with this representation. For example, the program in Figure 1(a) can be transformed into the layered network in Figure 1(c).

From the implementation point of view in PDGP programs are represented as arrays with the same topology as the grid, in which each cell contains a function label and the horizontal displacement of the nodes in the previous layer used as arguments for the function. It should be noted that, in order to exploit the possibilities offered by the use of "unexpressed" parts of code, functions or terminals are associated to *every* node in the grid, i.e. also to the nodes that are not directly or indirectly connected to the output. We call them *inactive nodes* or *introns*; the others are called *active nodes.*

This basic representation has been extended in various directions.[1] The first extension was to introduce vertical displacements to allow feed-forward connections between non-adjacent layers. With this extension any directed acyclic graph can be naturally represented within the grid of nodes, without requiring

---

[1] All these extensions require changes to the operators and interpreter used in PDGP. For the sake of brevity we do not discuss these changes in detail here.

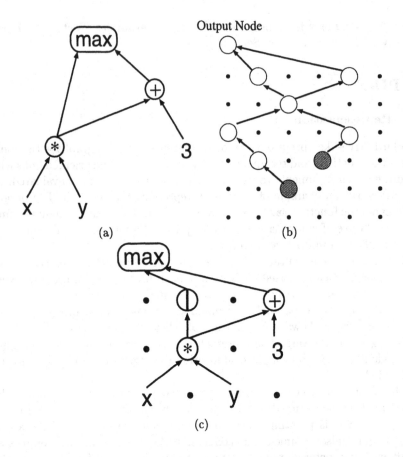

Fig. 1. (a) Graph-based representation of the expression `max(x * y, 3 + x * y)`; (b) Grid-based representation of graphs representing programs in PDGP (hollow circles represent functions, grey circles represent terminals); (c) Grid-based representation of the expression in (a).

the presence of pass-through nodes. The second extension has been to allow non-positive vertical displacements to represent graphs with backward connections (e.g. with cycles). The third extension has been to add labels to links. The link set can contain constants, variables, functions and macros. If the labels are numbers (e.g. ephemeral random numbers) this extension allows the direct development of neural networks. If the symbols of a language are used as labels for the links, it is possible to evolve finite state automata, transition nets, semantic nets, etc. If the links are functions or macros, then very complex behaviours can be realised, like the ones required to implement RTNs (see Section 3).

2.2 Genetic Operators

Several kinds of crossover, mutation and initialisation strategies can be defined for the basic representation used in PDGP and for its extensions. In the following we will only describe two forms of crossover (more details can be found in [17]).

The basic crossover operator of PDGP, which we call *Sub-graph Active-Active Node (SAAN) crossover*, is a generalisation to graphs of the crossover used in GP to recombine trees. SAAN crossover works as follows: 1) a random active node is selected in each parent (crossover point); 2) a sub-graph including all the active nodes which are used to compute the output value of the crossover point in the first parent is extracted; 3) the sub-graph is inserted in the second parent to generate the offspring (if the x coordinate of the insertion node in the second parent is not compatible with the width of the sub-graph, the sub-graph is wrapped around).[2]

The idea behind this form of crossover is that connected sub-graphs are functional units whose output is used by other functional units. Therefore, by replacing a sub-graph with another sub-graph, we tend to explore different ways of combining the functional units discovered during evolution. An example of SAAN crossover is shown in Figure 2.

Obviously, for SAAN crossover to work properly some care has to be taken to ensure that the depth of the sub-graph being inserted in the second parent is compatible with the maximum allowed depth, i.e. the number of rows in the grid. A simple way to do this is, for example, to select one of the two crossover points at random and choose the other with the coordinates of the first crossover point and the depth of the sub-graph in mind.

Several different forms of crossover can be defined by modifying SAAN crossover. The one which has given the best results until now is *Sub-Sub-graph Active-Active Node (SSAAN) Crossover*. In SSAAN crossover one crossover point is selected at random among the active nodes of each parent. Then, a complete sub-graph is extracted from the first parent (like in SAAN crossover) disregarding the problems possibly caused by its depth. If the depth of the sub-graph is too big for it to be copied into the second parent, the lowest nodes of the sub-graph are pruned to make it fit.[3] Figure 3 illustrates this process. In order for this type of crossover to work properly introns are essential, in particular the inactive terminals in the last row of the second parent (they have been explicitly shown in the figure to clarify the interaction between them and the functions and terminals in the sub-graph).

---

[2] The term "active-active" in the name of SAAN crossover derives from the fact that the crossover points selected in step 1 are active nodes for both parents. The term "sub-graph" refers to the fact that the nodes transferred from one parent to the other form a sub-graph.

[3] The term "sub-sub-graph" in the name of SSAAN crossover refers to the fact that, unlike in SAAN crossover, due to this pruning process the sub-graph actually transferred from one parent to the other may be a sub-graph of the sub-graph extracted from the first parent.

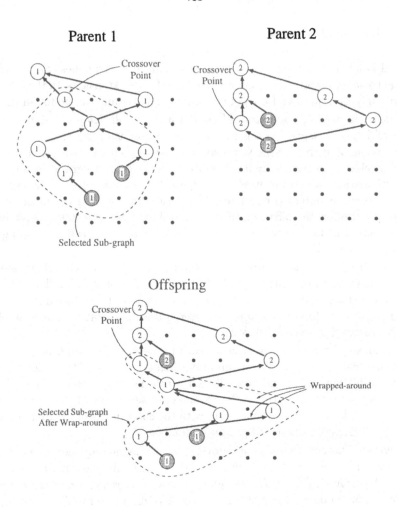

Fig. 2. Sub-graph active-active node (SAAN) crossover.

2.3 Interpreters

If no functions or terminals with side effects are used, it is possible to evaluate a PDGP program just like a feed-forward neural net, starting from the input-layer, which always contains only terminals, and proceeding layer after layer upwards until the output nodes are reached. However, the current PDGP interpreter does this differently, by using recursion and a hash table to store partial results and control information.

The interpreter, the procedure eval(program), can be thought of as having the structure outlined in Figure 4. eval starts the evaluation of a program from the output nodes and calls recursively the procedure microeval(node) to perform the evaluation of each of them. microeval checks to see if the value of a node is already known. If this is the case it returns the value stored in a hash table, otherwise it executes the corresponding terminal or function (calling itself

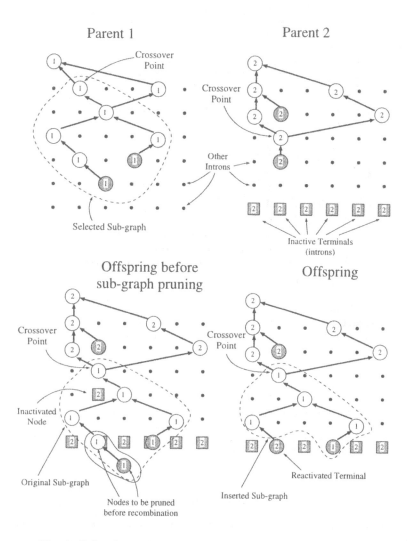

Fig. 3. Sub-sub-graph active-active node (SSAAN) crossover.

recursively to get the values for the arguments, if any).

This approach has the advantage of allowing the use of nodes with side effects[4] and of offering a total freedom in the connection topology of PDGP programs. Although the interpreter described above performs a sequential evaluation of programs, this process is inherently parallel. In fact, if we imagine each node to be a processor (with some memory to store its state and current value) and each link to be a communication channel, the evaluation of a program is equivalent to the parallel downward propagation of control messages requesting

---

[4] For brevity we do not describe here the changes needed to evaluate correctly nodes with side effects. More information on this is given in [17].

```
eval(program):
begin
    Reset the hash table NodeValue.
    For each node N in the  first layer (the output nodes) do
        Call the procedure microeval(N)
        Store the results in a temporary output vector Out
    Return(Out)
end
```

```
microeval(N):
begin
    If NodeValue(N) is "unknown" then
        If N is a function then
            For each node M connected (as an argument) to node N do
                Call the procedure microeval(M)
                Store the results in a temporary output vector Out
            Call the procedure N(Out)
            Store the result R in NodeVal(N)
            Return(R)
        elseif N is a macro then
            Call the procedure N(M1,...Mn) where M1...Mn  are the
                nodes connected (as  arguments) to node N
            Store the result R in NodeVal(N)
            Return(R)
        else /* N is a variable or a constant */
            Return(valof(N))
        endif
    else /* N has been already evaluated */
        Return(NodeVal(N))
    endif
end
```

Fig. 4. Pseudo-code for the interpreter of PDGP (in the absence of nodes with side effects).

a value followed by an upward propagation of return values. In this sense our interpreter can be seen as a parallel virtual machine capable of running data-driven dataflow programs [8] and PDGP as a method to evolve them.

3 Evolution of Natural Language Recognisers

While standard PDGP can evolve networks of any topology, in order to evolve RTNs for natural language processing we decided to use a technique unique to PDGP, called Automatically Defined Links (ADLs).

ADLs can be seen as the dual of Automatically Defined Functions (ADFs), which are sometimes used to evolve higher-level problem-specific primitives (parametrised subroutines) which improve the problem solving capabilities of GP [12]. ADFs are usually implemented using a special root node with multiple branches. One of the branches is interpreted as the value-returning (main) program, the others are interpreted as function definitions (the ADFs). The ADFs are part of the function set of the main program, which can therefore use them as individual-specific parametrised building blocks. Sometimes ADFs are part of the function set of other ADFs, too.

ADLs work in a very similar way. Basically, they are ADFs which, instead of being part of the function set of the main program, are part of the link set, and instead of being tree-like are graph-like subprograms. For example, if the label of a link is the symbol ADL0, the interpreter will "invoke" the parallel distributed program (in this case an RTN) corresponding to the first ADL. It should be noted that PDGP allows the use of both ADFs and ADLs.

For our experiments we generated 67 grammatical sentences and 181 ungrammatical ones, all with different syntactic structures, using the following grammar:

```
S   := NP VP
NP  := SNP | SNP PP
SNP := DET NOUN | DET ADJ NOUN | DET ADJ ADJ NOUN
PP  := PREP SNP
VP  := TVERB NP | IVERB
```

where S stands for sentence, NP for noun-phrase, VP for verb-phrase, SNP for simple NP, PP for prepositional phrase, DET for determiner, ADJ for adjective, TVERB for transitive verb, and IVERB for intransitive verb. The words in the examples were assigned a lexical category (like "noun", "verb", "adjective", etc.) so that the actual examples were sequences of lexical categories like DET ADJ NOUN TVERB DET ADJ NOUN PREP DET ADJ NOUN rather than "the little kitten frightened the big mouse with a sudden jump".

In the experiments we used a population with P=4,000 individuals, a maximum number of generations G=200, two ADLs, a regular 4 × 7 grid for the main RTN and two 4 × 5 grids for the ADLs. The recombination operator was a form of SSAAN crossover. The probability of crossover was 0.7, the probability of mutation was 0.1. The node in the top left corner of the main grid was considered to be the start node.

The following link set was used for both the main RTN and the two ADLs: [NOUN, TVERB, IVERB, PREP, DET, ADJ, ADL0, ADL1]. It includes the different kinds of lexical categories and the names of the two ADLs. The function set included the macros [N1 N2] which represent non-stop states in the RTN, with one and two outgoing links, respectively. When called they just check to see if the current symbol in the input sentence is present on one of their links. If this is the case they remove the symbol from the input stream and pass the control to the node connected to the link labelled with the matching symbol. If the label is an ADL, then control is passed to the corresponding RTN. If no link has the correct label, false is returned to the calling node. The same happens if no more symbols are available in the input stream.

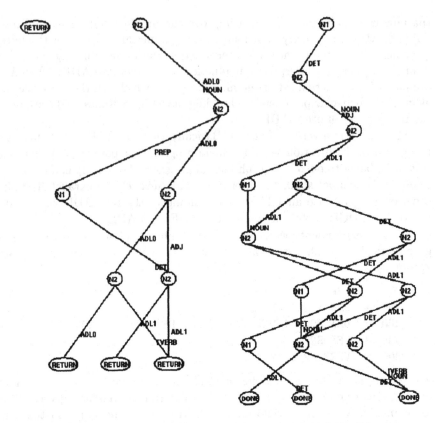

Fig. 5. Recursive transition network for recognising grammatical English sentences evolved by PDGP. The main RTN is shown on the right, ADL0 on the left and ADL1 in the centre.

The terminal set for the main RTN included only nodes of type DONE which behave like N1 nodes without output link except that if the end of the sentence is reached they return **true**. The terminal set for the ADLs included only nodes of type RETURN, which simply return **true**.

The fitness of an RTN was $f = 67 - n_s - 10^{-5} \times n_w$, where n_s is the number of sentences incorrectly classified and n_w is the number of words not read by the RTN in the sentences incorrectly classified. The value 67 corresponds to the number of grammatical sentences in the training set. The term $10^{-5} \times n_w$ has the function of smoothing the fitness landscape created by the term n_s. The coefficient 10^{-5} has been chosen so that $10^{-5} \times n_w < 1$. This ensures that evolution will always favour solutions with smaller n_s whatever the value of n_w.

```
> DET DET          !!> ADJ ADJ         !< PREP <false>     !< ADLO <true>
< DET <true>       !!< ADJ <true>      !> ADLO END         !> TVERB END
> ADJ NOUN         !!> ADL1 NOUN       !< ADLO <true>      !< TVERB <false>
< ADJ <false>      !!!> NOUN NOUN      !> ADJ END          !> ADLO END
> NOUN NOUN        !!!< NOUN <true>    !< ADJ <false>      !< ADLO <true>
< NOUN <true>      !!!> PREP TVERB     !> ADLO END         < ADL1 <true>
> DET PREP         !!!< PREP <false>   !< ADLO <true>      > DET END
< DET <false>      !!!> ADLO TVERB     !> TVERB END        < DET <false>
> ADL1 PREP        !!!< ADLO <true>    !< TVERB <false>    > ADL1 END
!> NOUN PREP       !!!> ADJ TVERB      !> ADLO END         !> NOUN END
!< NOUN <false>    !!!< ADJ <false>    !< ADLO <true>      !< NOUN <false>
!> ADLO PREP       !!!> ADLO TVERB     < ADL1 <true>       !> ADLO END
!< ADLO <true>     !!!< ADLO <true>    > DET END           !< ADLO <true>
!> PREP PREP       !!!> TVERB TVERB    < DET <false>       !> PREP END
!< PREP <true>     !!!< TVERB <true>   > ADL1 END          !< PREP <false>
!> DET DET         !!< ADL1 <true>     !> NOUN END         !> ADLO END
!< DET <true>      !< ADL1 <true>      !< NOUN <false>     !< ADLO <true>
!> ADL1 ADJ        < ADL1 <true>       !> ADLO END         !> ADJ END
!!> NOUN ADJ       > DET DET           !< ADLO <true>      !< ADJ <false>
!!< NOUN <false>   < DET <true>        !> PREP END         !> ADLO END
!!> ADLO ADJ       > DET NOUN          !< PREP <false>     !< ADLO <true>
!!< ADLO <true>    < DET <false>       !> ADLO END         !> TVERB END
!!> PREP ADJ       > ADL1 NOUN         !< ADLO <true>      !< TVERB <false>
!!< PREP <false>   !> NOUN NOUN        !> ADJ END          !> ADLO END
!!> ADLO ADJ       !< NOUN <true>      !< ADJ <false>      !< ADLO <true>
!!< ADLO <true>    !> PREP END         !> ADLO END         < ADL1 <true>
```

Fig. 6. Functions invoked by the parsing of the sentence "the kitten on the small mat frightened a mouse".

Figure 5 shows the 100%-correct solution found by PDGP after two unsuccessful runs. The performance of this RTN has been tested with 156 different grammatical sentences (all with different structure), which the system classified correctly, and 12066 different random ungrammatical sentence-types, of which the system correctly classified 12008.[5] This would suggests that the system has a sensitivity of 100% and a specificity of 99.5%. However, tests with 156 different slightly ungrammatical sentence-types (obtained by modifying one lexical category in a valid sentence) have revealed a lower specificity, 68.6%, which seems still quite good for a system 100% sensitive.

---

[5] The disparity between the number of grammatical and ungrammatical sentences is due to the impossibility of generating more grammatical sentence types with the grammar described above.

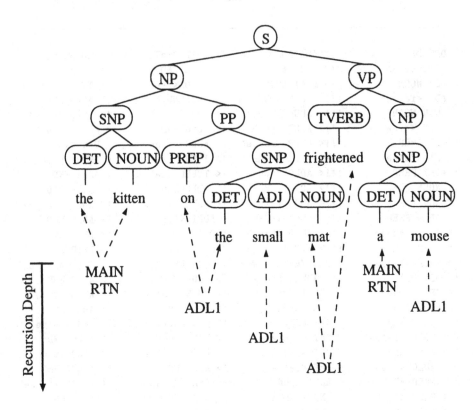

Fig. 7. Parse tree for the sentence "the kitten on the small mat frightened a mouse" and parts of the RTN responsible for its recognition (calls to ADL0 are not shown).

The recogniser works in a very complicated and unusual way, by using ADL0 to move from node to node without changing the symbol to be parsed, and ADL1 mainly to parse propositional phrases or fragments of them. For example, the output produced by the Pop-11 tracer when recognising the sentence "the kitten on the small mat frightened a mouse" is shown in Figure 6. The parse tree for this sentence is shown in Figure 7 together with the parts of the RTN responsible for the recognition of each word.

It should be noted that for very difficult problems which have never been solved before, at least with evolutionary techniques, such as the one considered in this paper or the evolution of electronic circuits (see for example [14]), it is usually impossible to perform more than a few runs (in this case 3). This is normally considered more than acceptable because, in these cases, the research focus is not assessing the performance of GP through the usual statistics used in the GP literature (like the computational effort, the percentage of successful runs, etc.), but rather showing how GP can reach new horizons and studying the characteristics of the solutions evolved.

4 Conclusions

In this paper we have presented PDGP, a new form of genetic programming which is suitable for the automatic discovery of parallel network-like programs, and applied it to the problem of inducing recognisers for natural language.

PDGP uses a grid-based representation of programs which allowed us to develop efficient genetic operators. The programs developed by PDGP are fine-grained, but the representation used is also suitable for the development of medium-grained parallel programs via the use of automatically defined functions and automatically defined links: a new technique unique to PDGP.

The experimental results with the evolution of recursive transition nets for natural language recognition obtained using this technique are very promising and show how evolutionary computation with the use of little prior knowledge can now tackle higher and higher level problems usually considered to require special AI techniques.

It should be noted that ADLs are just one of the new representational possibilities opened by PDGP. PDGP is an efficient paradigm to optimise general graphs (with or without cycles, possibly recursively nested via ADFs and ADLs, with or without labelled links, with or without directions, etc.). These graphs need not be interpreted as programs. They can be interpreted as engineering designs, semantic nets, neural networks, etc. Also, cellular encoding can naturally be extended to PDGP, thus creating a very large set of new possibilities.

Acknowledgements

The author wishes to thank Bill Langdon and the other members of the Evolutionary and Emergent Behaviour Intelligence and Computation (EEBIC) group for useful discussions and comments. The anonymous reviewers are also thanked for their invaluable help in improving this paper.

References

1. *Late Breaking Papers at the Genetic Programming 1996 Conference*, Stanford University, July 1996. Stanford Bookstore.
2. David Andre, Forrest H. Bennett III, and John R. Koza. Discovery by genetic programming of a cellular automata rule that is better than any known rule for the majority classification problem. In John R. Koza, David E. Goldberg, David B. Fogel, and Rick L. Riolo, editors, *Genetic Programming 1996: Proceedings of the First Annual Conference*, pages 3–11, Stanford University, CA, USA, 28–31 July 1996. MIT Press.
3. Forrest H. Bennett III. Automatic creation of an efficient multi-agent architecture using genetic programming with architecture-altering operations. In John R. Koza, David E. Goldberg, David B. Fogel, and Rick L. Riolo, editors, *Genetic Programming 1996: Proceedings of the First Annual Conference*, pages 30–38, Stanford University, CA, USA, 28–31 July 1996. MIT Press.

4. Scott Brave. Evolving deterministic finite automata using cellular encoding. In John R. Koza, David E. Goldberg, David B. Fogel, and Rick L. Riolo, editors, *Genetic Programming 1996: Proceedings of the First Annual Conference*, pages 39–44, Stanford University, CA, USA, 28–31 July 1996. MIT Press.

5. Ted E. Dunning and Mark W. Davis. Evolutionary algorithms for natural language processing. In John R. Koza, editor, *Late Breaking Papers at the Genetic Programming 1996 Conference Stanford University July 28-31, 1996*, pages 16–23, Stanford University, CA, USA, 28–31 July 1996. Stanford Bookstore.

6. F Gruau and D. Whitley. Adding learning to the cellular development process: a comparative study. *Evolutionary Computation*, 1(3):213–233, 1993.

7. Frederic Gruau. Genetic micro programming of neural networks. In Kenneth E. Kinnear, Jr., editor, *Advances in Genetic Programming*, chapter 24, pages 495–518. MIT Press, 1994.

8. R. Jagannathan. Dataflow models. In E. Y. Zomaya, editor, *Parallel and Distributed Computing Handbook*. McGrow-Hill, 1995.

9. K. E. Kinnear, Jr., editor. *Advances in Genetic Programming*. MIT Press, 1994.

10. J. R. Koza, D. E. Goldberg, D. B. Fogel, and R. L. Riolo, editors. *Proceedings of the First International Conference on Genetic Programming*, Stenford University, July 1996. MIT Press.

11. John R. Koza. *Genetic Programming: On the Programming of Computers by Means of Natural Selection*. MIT Press, 1992.

12. John R. Koza. *Genetic Programming II: Automatic Discovery of Reusable Programs*. MIT Pres, Cambridge, Massachusetts, 1994.

13. John R. Koza, David Andre, Forrest H. Bennett III, and Martin A. Keane. Use of automatically defined functions and architecture-altering operations in automated circuit synthesis using genetic programming. In John R. Koza, David E. Goldberg, David B. Fogel, and Rick L. Riolo, editors, *Genetic Programming 1996: Proceedings of the First Annual Conference*, pages 132–149, Stanford University, CA, USA, 28–31 July 1996. MIT Press.

14. John R. Koza, Forest H. Bennett III, Jason Lohn, Frank Dunlap, Martin A. Keane, and David Andre. Use of architecture-altering operations to dynamically adapt a three-way analog source identification circuit to accommodate a new source. In John R. Koza, Kalyanmoy Deb, Marco Dorigo, David B. Fogel, Max Garzon, Hitoshi Iba, and Rick L. Riolo, editors, *Genetic Programming 1997: Proceedings of the Second Annual Conference*, Stanford University, CA, USA, 13-16 July 1997. Morgan Kaufmann.

15. John R. Koza, Forrest H. Bennett III, David Andre, and Martin A. Keane. Automated WYWIWYG design of both the topology and component values of electrical circuits using genetic programming. In John R. Koza, David E. Goldberg, David B. Fogel, and Rick L. Riolo, editors, *Genetic Programming 1996: Proceedings of the First Annual Conference*, pages 123–131, Stanford University, CA, USA, 28–31 July 1996. MIT Press.

16. Thomas R. Osborn, Adib Charif, Ricardo Lamas, and Eugene Dubossarsky. Genetic logic programming. In *1995 IEEE Conference on Evolutionary Computation*, volume 2, page 728, Perth, Australia, 29 November - 1 December 1995. IEEE Press.

17. R. Poli. Parallel distributed genetic programming. Technical Report CSRP-96-15, School of Computer Science, The University of Birmingham, September 1996.

18. Riccardo Poli. Genetic programming for image analysis. In John R. Koza, David E. Goldberg, David B. Fogel, and Rick L. Riolo, editors, *Genetic Programming 1996: Proceedings of the First Annual Conference*, pages 363–368, Stanford University, CA, USA, 28–31 July 1996. MIT Press.

19. Riccardo Poli. Some steps towards a form of parallel distributed genetic programming. In *Proceedings of the First On-line Workshop on Soft Computing*, August 1996.

20. Riccardo Poli. Discovery of symbolic, neuro-symbolic and neural networks with parallel distributed genetic programming. In *3rd International Conference on Artificial Neural Networks and Genetic Algorithms, ICANNGA'97*, 1997.

21. Stuart C. Shapiro. *Encyclopedia of Artificial Intelligence*. Wiley, New York, second edition, 1992.

22. Eric V. Siegel. Competitively evolving decision trees against fixed training cases for natural language processing. In Kenneth E. Kinnear, Jr., editor, *Advances in Genetic Programming*, chapter 19, pages 409–423. MIT Press, 1994.

23. Astro Teller and Manuela Veloso. PADO: Learning tree structured algorithms for orchestration into an object recognition system. Technical Report CMU-CS-95-101, Department of Computer Science, Carnegie Mellon University, Pittsburgh, PA, USA, 1995.

24. Paul Walsh and Conor Ryan. Paragen: A novel technique for the autoparallelisation of sequential programs using genetic programming. In John R. Koza, David E. Goldberg, David B. Fogel, and Rick L. Riolo, editors, *Genetic Programming 1996: Proceedings of the First Annual Conference*, pages 406–409, Stanford University, CA, USA, 28–31 July 1996. MIT Press.

Progress in Evolutionary Scheduling

Scheduling Planned Maintenance of the South Wales Region of the National Grid

W. B. Langdon*

School of Computer Science, University of Birmingham,
Birmingham B15 2TT, UK

Abstract. The maintenance of the high voltage electricity transmission network in England and Wales (the National Grid) is planned so as to minimise costs taking into account:

1. location and size of demand for electricity,
2. generator capacities and availabilities,
3. electricity carrying capacity of the remainder of the network, i.e. that part not undergoing maintenance,
4. resilience of the network to faults (contingencies).

This complex optimisation and scheduling problem is currently performed manually (albeit with some computerised assistance). This paper reports recent work aiming to automatically generate low cost schedules using genetic algorithms (GA) using the South Wales region as a demonstration network.

The combination of a "greedy optimiser" with a permutation GA, which has been demonstrated on a small network, was successfully applied to the South Wales network.

1 Introduction

In England and Wales electrical power is transmitted by a high voltage electricity transmission network which is highly interconnected and carries large power flows. It is owned and operated by The National Grid Company plc. (NGC) who maintain it and wish to ensure its maintenance is performed at least cost, consistent with plant safety and security of supply.

There are many components in the cost of planned maintenance. The largest is the cost of replacement electricity generation, which occurs when maintenance of the network prevents a cheap generator from running so requiring a more expensive generator to be run in its place.

The task of planning maintenance is a complex constrained optimisation scheduling problem. The schedule is constrained to ensure that all plant remains within its capacity and the cost of replacement generation, throughout the duration of the plan is minimised. At present maintenance schedules are initially produced manually by NGC's Planning Engineers (who use computerised viability checks and optimisation on the schedule after it has been produced). This

---

[1] *email:* W.B.Langdon@cs.bham.ac.uk

paper describes work by NGC's Technology and Science Laboratories and University College London which investigates the feasibility of generating practical and economic maintenance schedules using genetic algorithms (GAs) [Hol92].

Earlier work [Lan95] investigated creating electrical network maintenance plans using Genetic Algorithms on a demonstration four node test problem devised by NGC. [Lan95] showed the combination of a Genetic Algorithm using an order or permutation chromosome combined with hand coded "Greedy" schedulers can readily produce an optimal schedule for this problem.

In this paper we report work showing the combination of GAs and greedy optimisers described in [Lan95] being applied to the South Wales region of the National Grid. Other work on the South Wales problem using Genetic Programming is reported in [LT97].

Section 2 describes the South Wales region high voltage network. Sections 3 and 4 describe the fitness functions and introduce the hybrid GA used in Sects. 5 and 6 in which low cost schedules are produced, firstly without the network resilience requirement and secondly when including it. Our conclusions, in Sect. 7, are followed in Sect. 8 by our suggestions for future work in particular applying our approach to the larger national network. Details of the GA parameters used are given in Appendix A.

2   The South Wales Region of the UK Electricity Network

The South Wales region of the UK electricity network carries power at 275K Volts and 400K Volts between electricity generators and regional electricity distribution companies and major industrial consumers. The region covers the major cites of Swansea, Cardiff, Newport and Bristol, steel works and the surrounding towns and rural areas (see Fig. 1). The major sources of electricity are infeeds from the English Midlands, coal fired generation at Aberthaw, nuclear generation at Oldbury and oil fired generation at Pembroke. The two infeeds are shown as the large black circles labelled WALHA1 and MELKA1 in Figs. 1 and 7. Black circles represent sources of electrical power, grey circles represent locations of demand and white circles represent other connections in the network, possibly between lines operating at different voltages. Oldbury and Pembroke are shown in grey as demand on these nodes in the network in 1995 was typically greater than the electricity generated by the power station. High capacity lines are shown with thick lines.

Demand for electricity varies considerably throughout the year. There is basic seasonal variation with a peak in midwinter caused by additional demand for heating and lighting. On top of this there are short periods of reduced demand caused by the three traditional holiday seasons in South Wales (See Fig. 2). The four consecutive weeks with the lowest demand (weeks 18 to 21) are highlighted with arrows in Figs. 2 and 6.

The demand for electricity must be met exactly by generation. Figure 3 shows the predicted amount of power to be imported into the South Wales region (two upper lines) and the output of each generator in the region throughout the year

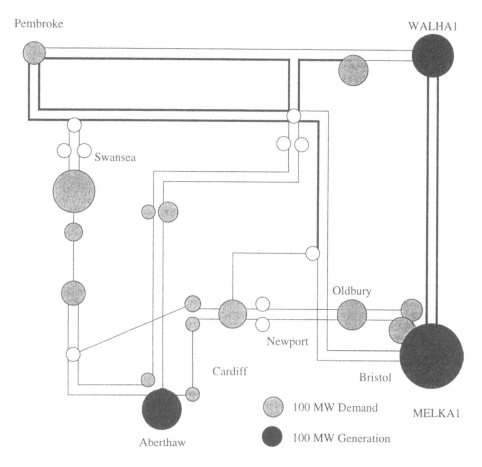

Fig. 1. South Wales Region High Voltage Network

(shown with symbols). The predicted generation pattern is pre-calculated on the basis of nation wide optimisations. For the purposes of planning network maintenance it can be taken as fixed. From Fig. 3 we can see that most of the generators within the region are fairly static. I.e. they run either at full output or not at all most of the time. Most of the variation in demand is met by changing the amount of electricity supplied from outside the region (the upper two lines on Fig. 3).

The representation of the electricity network used in these experiments is based upon the engineering data available for the physical network; however a number of simplifications have to be made. Firstly the regional network has been treated as an isolated network; its connections to the rest of the network have been modelled by two sources of generation connected by a pair of low impedance high capacity conductors. Secondly the physical network contains short spurs run to localised load points such as steel works. These "T" points have been simplified (e.g. by inserting nodes into the simulation) so all conductors connect two nodes.

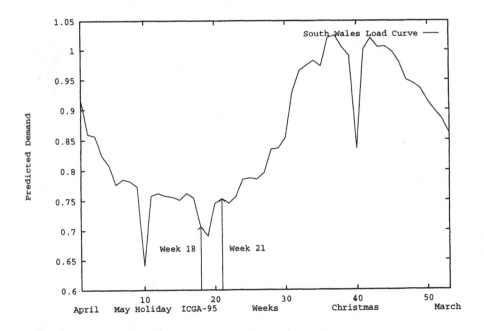

Fig. 2. Predicted Demand in South Wales Region

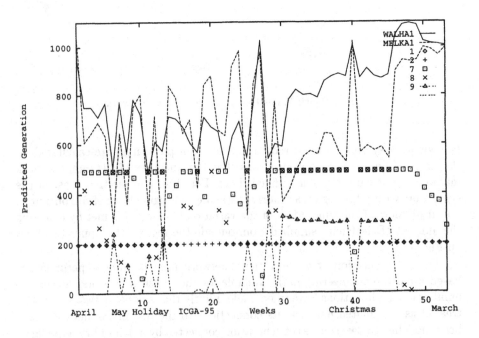

Fig. 3. Predicted Generation in South Wales Region

The industry standard DC load flow approximation is used to calculate power flows through the network.

In the experiments reported in this paper the maintenance planning problem for the South Wales region has been made deliberately more difficult than the true requirement because we are not considering the whole network. In these experiments:

1. All lines must be maintained during the 53 week plan (1995 had 53 weeks rather than 52). Typically about one third to a half of the lines are maintained in any one year.
2. All maintenance takes four weeks. Typically scheduled outage of a line is between a week and $1\frac{1}{2}$ months.
3. We consider two versions of the problem. In the first the requirement that the network must remain secure against faults in the network was ignored. To compensate for this, the problem was artificially increased in difficulty by reducing the capacity of all lines by 50%. When satisfactory solutions to this version of the problem have been evolved, we then considered evolving maintenance plans which contain a degree of network resilience (cf. Sect. 6). In this second problem the true conductor rating were used.

Considering potential network faults is highly CPU intensive and so the Genetic programming approach described in [LT97] did not attempt to solve the second version of the South Wales problem.

3 The Fitness Function

The fitness function used by both GA and genetic programming (GP) approaches to scheduling maintenance in the South Wales region is based on that used in the four node problem. To summarise;

1. The demand for electricity for the coming year is predicted (cf. Fig. 2).
2. The cheapest electrical generators are scheduled to meet this demand (cf. Fig. 3).
   To optimise costs across the whole network, these two steps are performed nationally. Imbalances between demand and generation in each region result in power transfers between regions and in the case of the South Wales region for this year, large inward power flows were planned (cf. the top two curves in Fig. 3).
3. The fitness function includes a maintenance benefit for each maintenance task completed (i.e. for each line maintained for at least four consecutive weeks).
4. Large penalties are also included to punish schedules which isolate nodes which directly supply demand or generation. When nodes are isolated or the network is split, the penalty for each node in the disconnected part of the network are summed and then the usual loadflow and fitness calculations are performed on the remaining connected network (i.e. the connected network

containing the most nodes). Unlike the four node problem, in the South Wales problem there is no separate penalty for splitting the network.

5. If maintenance were to cause power flows to be redistributed in such a way as to overload part of the network it would be necessary to re-arrange the plan to avoid this. In practice this may entail changes to the pattern of electricity generation. As we started with the cheapest generators, this will be more expensive. In practice this is the major cost of planned network maintenance. The fitness function models this cost with a penalty given by the amount each line is overloaded.

The cost of trial schedules is calculated in terms of energy (MW weeks or $6.048 \; 10^{11}$ J). While a cash value could be ascribed to trial schedules, this is not required by the GA, which needs only to compare relative performance of candidate schedules.

$$
\text{cost} = \sum_{\text{target}} K \times (1 - \text{maintenance\_scheduled})
$$

$$
+ \sum_{\text{weeks}} \left(+ \begin{array}{l} S \times \sum_{\text{isolated\_nodes}} |\text{ demand} \vee \text{generation }| \\[2ex] \sum_{\text{remaining\_lines}} \begin{array}{l} \textbf{If } |\text{flow}| > \text{rating } \textbf{then} \\ |\text{flow}| - \text{rating} \end{array} \end{array} \right) \quad . \quad (1)
$$

Equation (1) gives the fitness function used in Sect. 5, i.e. excluding consideration of contingencies (network faults). In (1) the first summation is over all target maintenance and is responsible for calculating the maintenance benefit, item 3. above. (NB the trial plan's cost is increased by K for each line not maintained). The second outer summation being over each week of the maintenance plan; the first inner one, being over all isolated nodes, and the second, over all lines in the network. The two inner summations are responsible for calculating the node isolation and line over loading penalties respectively, items 4. and 5. above.

Where contingencies are to be considered (as in Sect. 6), the second part of (1) is extended to include a penalty for each contingency which is calculated as if the contingency had occurred but weighted by an arbitrary scaling factor based upon the severity of the contingency (cf. (2)). During the contingency part of the fitness calculation, i.e. the third outer summation in (2), the post fault rating on all the conductors is used. This is about 17% higher for each conductor than the corresponding rating used in the first part. Apart from this calculating the node isolation and line over loading penalties (items 4. and 5. above) for each contingency is the same as for the base case. It can be appreciated that, where there are many contingencies to consider, this adds considerably to the time taken to evaluate each schedule's fitness.

$$\text{cost} = \sum_{\text{target}} K \times (1 - \text{maintenance\_scheduled})$$

$$+ \quad \sum_{\text{weeks}} \left(\begin{array}{l} S \times \displaystyle\sum_{\text{isolated\_nodes}} \mid \text{demand} \vee \text{generation} \mid \\[1em] + \\[0.5em] \quad \displaystyle\sum_{\text{remaining\_lines}} \begin{array}{l} \textbf{If } \mid\text{flow}\mid > \text{rating } \textbf{then} \\ \quad \mid\text{flow}\mid - \text{rating} \end{array} \end{array} \right) .$$

$$+ \quad \sum_{\text{contingency}} \text{scaling} \times \sum_{\text{weeks}} \left(\begin{array}{l} S \times \displaystyle\sum_{\text{isolated\_nodes}} \mid \text{demand} \vee \text{generation} \mid \\[1em] + \\[0.5em] \quad \displaystyle\sum_{\text{remaining\_lines}} \begin{array}{l} \textbf{If } \mid\text{flow}\mid > \text{rating}_{+17\%} \textbf{then} \\ \quad \mid\text{flow}\mid - \text{rating}_{+17\%} \end{array} \end{array} \right)$$

$$(2)$$

The scaling constants, K and S need to be chosen so that the maintenance benefit dominates. For the four node problem this was done by calculating the line costs associated with many feasible schedules [Lan95, page 140]. This gave the highest feasible line cost as being 3870MW weeks. K was set to 4,000MW weeks, so that any schedule which failed to maintain even one line would have a higher cost that the worst "reasonable" schedule which maintains them all.

S was chosen so that isolating any node would have a high cost. In the four node problem the lowest demand/generation on a node is 800MW. Setting S to five ensures that any schedule which isolates any node will have a higher cost than the worst "reasonable" schedule (800MW × 5weeks = 4,000MW weeks).

For the South Wales problem the same values of K and S as the four node system where used. In general the balance between the costs of maintenance and the (typically unknown) costs of not maintaining a plant item is complex. Such choices are beyond the scope of this work, however [Gor95] verified these values are reasonable for the South Wales region.

The South Wales network is obviously considerably larger and more complex than the four node network. Correspondingly, both the time to perform network connectivity checks and to calculate powerflows within the network are greatly increased. Since both are fundamental parts of some greedy schedulers and of the fitness function, the run time of the South Wales problem is far greater than that of the four node problem.

Analysis indicates that performing the network connectivity checks can be a bottle neck. A significant reduction in run time was achieved by replacing the original algorithm. Additional reduction was achieved by maintaining a cache of previously evaluated network connectivities. It is felt that caching techniques could produce worthwhile reductions in run time if also applied to the loadflow calculation.

4 Greedy Optimisers

The approach taken to solving the power transmission network maintenance scheduling problem has been to split the problem in two; a GA and a "greedy optimiser". The greedy optimiser is presented with a list of work to be done (i.e. lines to be maintained) by the GA. It schedules those lines one at a time, in the order presented by the GA, using some problem dependent heuristic. Figure 4 shows this schematically.

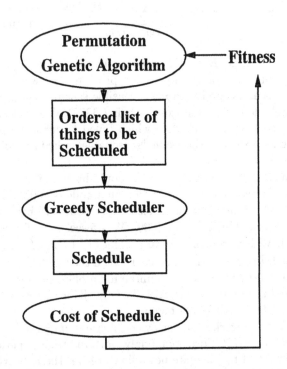

Fig. 4. Hybrid GA and "greedy optimiser"

This approach of hybridising a GA with a problem specific heuristic has been widely studied. Davis [Dav91] for example firmly advocates using hybrid GAs when attempting to solve difficult real world problems. Hybrid GAs, of various sorts, have been used on a number of scheduling problems (e.g. flight crew scheduling [Lev94], task scheduling [Sys91] and job-shop and open-shop scheduling [FRC93] [FRC94] [YN92]). Potential problems with using other approaches on our problem were discussed in [Lan95].

5   South Wales Problem without Contingencies

5.1   Greedy Scheduler

The greedy scheduler used is based upon the minimum increase cost greedy
scheduler that solved the four node problem (Heuristic 4 in [Lan95]) but modified
to take into account maintenance takes four weeks rather than one. The greedy
scheduler (cf. Fig. 4) is presented with each line in the order determined by the
GA chromosome. For each line it determines which four weeks will cause the
overall cost of the schedule to increase the least (or decrease the most). The line
is then scheduled in those four weeks and the next line is processed. If there is a
tie, the earliest weeks are used. NB this optimiser makes its choice by studying
the effect each maintenance outage would have on the network, rather than using
only the existing network information. I.e. it performs a limited look ahead.

5.2   Results

Figure 5 contains the least cost schedule produced using the minimum increase
cost greedy scheduler using the same GA that was used to solve the four node
problem with the same population size (cf. Appendix A).

   Not only is this a low cost schedule (it has a cost of 616MW weeks) but
it is a good one in that it contains several "sensible" decisions, for example
performing maintenance on lines and the transformers which supply them simultaneously. For example line Swansea L9 and transformer Swansea T8, which
supply Swansea, are both maintained in weeks 4–7. Exactly the opposite happens with the two transformers at MELKA1 because they are both directly
connected to demand and so it is important that the line and the transformer
are not simultaneously maintained.

5.3   Discussion

While still better performance could be obtained using modified greedy schedulers, these results were sufficiently encouraging as to warrant including the fault
resilience requirements by explicitly including contingencies. This is described in
the next section.

   The run time of these experiments are consistent with the $O(\text{weeks} \times \text{lines} \times \text{nodes}^3)$ prediction made in [Lan95, page 148]. It is encouraging that reasonable
results have been obtained using essentially the same technique as was used on
the four node problem, despite this problem's search space being $2.1 \ 10^{60}$ fold
larger.

   Another heuristic, which produced improved results, replaced the line over-
loading component of the minimum increase in cost greedy scheduler with the
square of the powerflow. It is felt this produced improved performance because
in the case of the South Wales network, unlike the four node problem, many conductors are not near their ratings and so the minimum increase in cost heuristic
is relatively independent of changes to their powerflows. By using power flow

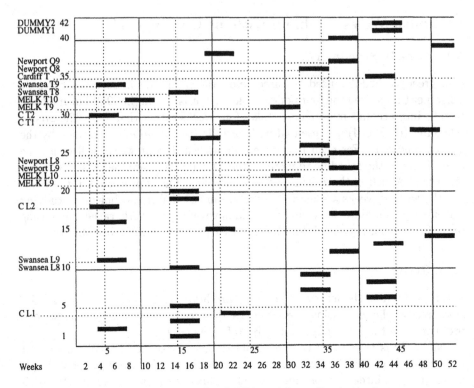

Fig. 5. Schedule for South Wales without Contingencies

squared, the greedy scheduler becomes more sensitive to changes but particularly to changes in lines near their capacities.

On this version of the South Wales problem and using a population seeded with the same (and other) heuristics, GP was able to find schedules with a cost of 388MW weeks albeit with far greater resources (e.g. the population was 1000) [LT97].

[Gor95] used a linear chromosome with non-binary alleles [Ros94] to solve the four node problem but was less successful on this version of the South Wales problem.

6 South Wales Problem with Contingencies

In the experiments reported in this section all 52 contingencies (i.e. potential faults which prevent a line carrying power) were included. This greatly increases run time. The 52 contingencies comprise; 42 single fault contingencies (one for each line in the network and twelve double fault contingencies (for cases where lines run physically close together, often supported by the same towers). The scaling factors for the single fault contingencies were 0.1, with 0.25 for the double fault contingencies, except the most severe where it was 1.0.

6.1 Greedy Scheduler

A number of "Greedy Schedulers" were tried with the permutation GA (described in Appendix A). Satisfactory schedules where obtained when considering all 52 contingencies, using a two stage heuristic.

The choice of which four weeks to schedule a line's maintenance in is performed in two stages. In the first stage, we start with the maintenance schedule constructed so far and calculate the effect on network connectivity of scheduling the line in each of the 53 available weeks if each of the possible contingencies occurred simultaneously with the line's maintenance. (NB one contingency at a time). To reduce computational load, at this stage we only consider if the result would cause disconnections, i.e. we ignore line loadings. In many cases none of the contingencies would isolate nodes with demand or generation attached to them and all weeks are passed onto the second stage. However as the maintenance schedule becomes more complete there will be cases where one or more contingencies would cause disconnection in one or more weeks. Here the sum over all contingencies, of the load disconnected, weighted by the each contingency's' scaling factor, is calculated and only those weeks where it is a minimum are passed to the second stage.

In the second stage the modified version of the minimum increase in line cost heuristic, used in Sect. 5.1, is used to select the weeks in which to perform maintenance. NB the second stage can only choose from weeks passed to it by the contingency pre-pass. By considering contingencies only in the first stage, run time is kept manageable.

6.2 Results

The least cost schedule found in generation 18 after 78 minutes of one run of QGAME on a 50 MHz SUN server 690MP is shown in Fig. 6. The cost of the schedule is dominated by the contingency costs incurred in weeks 18–21. Four of the six lines maintained during these four weeks will always incur contingency costs, because once taken out of service nearby demand nodes are only connected to the remaining network by a single line (see Fig. 7). By placing these in weeks 18–21 their contingency costs have been minimised as these weeks correspond to minimum demand.

6.3 Discussion

While the schedules found do represent low cost solutions to the problem presented to the GA, unlike in the four node problem, they may not be optimal. With problems of this complexity, low cost solutions are all that can reasonably be expected. Computer generated schedules have been presented to NGC's planning engineers, while expressing some concerns, they confirm they are reasonable.

The two principle concerns were; firstly the GA appears to have been concerned to optimise the contingency costs associated with small demand nodes

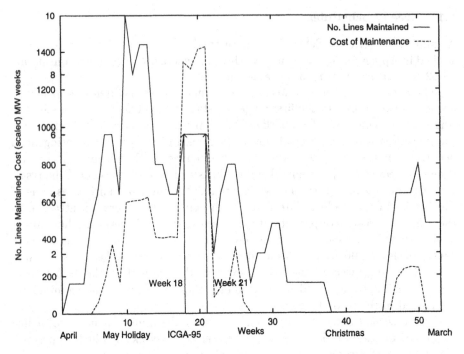

Fig. 6. South Wales with Contingencies Maintenance Schedule

with only two connections. Since contingency costs associated with them are inevitable, currently planning engineers do not concentrate on minor improvements to them.

The greedy scheduler approach, schedules each line one at a time. There is a concern that there may be circumstances where the optimal solution requires two lines to be maintained at times which do not correspond to the local best for either of them. For example suppose it is planned that a generator will be run at half power for eight weeks and that it is connected by several lines, two of which are to be maintained. It might make sense to maintain them one after the other (so the generator is always connected by as many lines as possible) during the time the generator will be at low power. However the greedy scheduler could decide the best time to schedule one of them (looking just at the best for that line) was in the third of these eight weeks. This would prevent the second line being maintained during the low power interval, so preventing the optimal schedule from being found. While this special case does not arise in the South Wales problem it, or similar problems, may occur in the complete network. We feel that the greedy scheduler approach is sufficiently flexible that it could, it if necessary, be readily extended to include consideration of special cases. However the approach is deliberately general purpose with a view to discovering novel schedules which an approach constrained by existing engineering wisdom might miss.

Since this work began NGC have continued to extend the computer tools

Pembroke

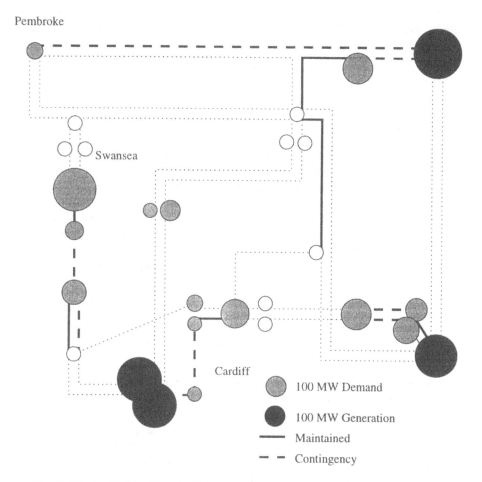

100 MW Demand

100 MW Generation

——— Maintained

— — Contingency

Fig. 7. Weeks 18–21 of South Wales with Contingencies Maintenance Schedule

available to assist their planning engineers. Some new tools have assisted in the discovery of improvements to existing schedules but production of new maintenance planning schedules remains a very difficult problem whose solution requires very skilled operators but offers financial savings.

7 Conclusions

We have described the complex real world problem of scheduling preventive maintenance of a very large electricity transmission network. The combination of a GA and hand coded heuristic has been shown to produce reasonable schedules for a real regional power network when including consideration of network robustness to single and double failures. However consideration of such contingencies considerably increases run time and so the production of schedules for the entire national network has not yet been demonstrated.

The time taken to perform GA fitness evaluations and with it program run time, grows rapidly with problem size and number of potential failures that must be considered. It is anticipated that running on parallel machines will be required to solve the national problem using a GA or GP approach. However there are a number of techniques which could be used to contain run time.

8 Future Work

Having evolved satisfactory fault tolerant schedules for the preventive mainte-nance of the South Wales high voltage regional electricity transmission network. (Indeed we have tackled a version of the problem which schedules maintenance for many more plant items than are typically required). We now consider how this technique will scale up to larger problems, such as the NGC network. Cur-rently the problem is solved manually (an automatic single pass optimiser can now be run to optimise manually chosen critical components of the manually produced schedule). This paper has already demonstrated the feasibility of us-ing genetic algorithms to automate this for a regional network of 42 nodes, albeit with a simple cost function. In this last section we discuss using our GA on the complete NGC network.

The NGC network contains about ten times as many nodes and lines as the South Wales network and requires approximately two to three times as many maintenance activities to be scheduled than we considered in this paper. In this paper we considered all possible single and double fault contingencies. In practice run time consideration mean a study of the complete network is unlikely to consider all possible contingencies.

There are two aspects to scaling up; how difficult it is to search the larger search space and how the run time taken by the schedule builder and fitness evaluation scales. A large component of run time is the DC load flow calculation, which assuming the number of nodes $\approx$ number of lines, is $O(n^3)$. In practice we expect to do rather better because:

1. We expect to use improved algorithms for DC load flow calculation, such as the sparse matrix linear programming package SDRS2 and the Woodbury Formula [PTVF92].
2. The may be considerable scope for the uses of caches of previously calcu-lated partial results which can be reused on different members of the GA populations,
3. The code has not been optimised as yet.

But even so, considerable growth in the fitness evaluation time must be antici-pated, as the size of the problem is increased and parallel execution on multiple CPUs is anticipated.

We can also calculate how the size of the search space grows with problem size but assessing how difficult it will be to search is more problematic. The size of the complete search space grows exponentially with the number of lines to be maintained, being $2^{maintenance \times weeks}$. However this includes many solutions that we are not interested in, e.g. we are not interested in solutions which take a line out of service when all its maintenance has already been done.

If we are prepared to consider only solutions which maintain each line in a single operation for exactly the required time then the number of such solutions is:

$$(1 + \text{weeks} - \text{weeks to do maintenance})^{\text{maintenance}} \qquad (3)$$

Although a big number this is considerably less than $2^{\text{maintenance} \times \text{weeks}}$. For the South Wales problem the ratio between them is $1.3 \ 10^{-586}$ and this difference grows exponentially as the problem increases in size to $1.3 \ 10^{-1812}$ for the NGC problem.

The greedy optimisers use their problem specific knowledge to further reduce the search space. They consider (maintenance)! points in the smaller search space referred to in (3). Once again this is a considerable reduction. For the South Wales problem it is a $2.8 \ 10^{-21}$ fold reduction and this difference grows rapidly as the problem increases in size. However there is a danger in all such heuristics that by considering some solutions rather than all, the optimum solution may be missed.

There is no guarantee that NGC problem's fitness landscapes will be similar to that of the regional network. However we can gain some reassurance from the fact that the GA which optimally solved the four node problem also produced good solutions to the bigger regional problem. This gives us some comfort as we increase the problem size again.

A Using QGAME

QGAME [RT94] was used in all the GA experiments on grid scheduling problems. In every case the genetic algorithm contain a single population of 20 individuals, each contained a chromosome which consisted of a permutation of the lines to be scheduled. Permutation crossover and the default parameters were used (see Table 1).

QGAME provides one permutation crossover, it is similar to PMX [Gol89]. The permutation crossover firstly selects one crossover point at random. One point crossover [Gol89, page 12] is performed on the two parents to create two new chromosomes. Even though both parents are permutations of the first n integers (where n is the length of the chromosomes, i.e. the number of tasks to be scheduled), at this point the two offspring may not be valid permutations. I.e. if the parents are different, one point crossover may create offspring which contain two copies of some integers and are missing others. Accordingly a second stage fix up stage is used to ensure both offspring are valid. This checks none of the genes to the right of the crossover point are already present in the left part. If an integer was inherited from both parents, the locus in the right hand part of the chromosome containing it is overwritten by the value in the same locus in the parent from which the left hand part was inherited. If both parents are valid permutations, this ensures both offspring will be valid permutations. NB both offspring inherit all of their left hand genes from one parent and the genes to

Table 1. QGAME genetic algorithm parameters

Name	Value	Meaning
Problem_type	Combinatorial	Combinatorial crossover and mutation operators are used on the chromosome, which holds a permutation of the lines to be scheduled
GeneType	G_INT	Line identification number
POPULATION	20	Size of population
CHROM_LEN	42	Forty two genes (lines) per chromosome
OptimiseType	Minimise	Search for lowest fitness value
PCROSS	0.6	On average each generation 0.6×POPULATION/2 pairs of individuals are replaced by two new ones created by crossing them over
PMUT	0.03	On average each generation 0.03×POPULATION individuals are mutated by swapping two genes
Selection	Truncated	The best TRUNCATION×POPULATION of the population are passed directly to the next generation. The remainder are selected using normalised fitness proportionate (roulette wheel) selection. Fitness is linearly rescaled so the best in the population has a normalised fitness of 1.0 and the worst zero
TRUNCATION	0.5	The best 50% passed directly to the next generation
POOLS	1	Single population
CHROMS	1	Individuals use single chromosome
MAX_GEN	100	Run for 100 generations

the right of the crossover point come primarily from the other parent but some may be inherited from the first parent. (The greedy scheduler starts processing the chromosome from its left hand end).

Acknowledgements

This work was carried out at UCL Computer Science department and funded by the EPSRC and The National Grid Company plc.

I would like to thank Tom Westerdale, and Mauro Manela for their critisims and ideas; John Macqueen and Maurice Dunnett who introduced me to the NGC maintenance planning problem. Mike Calviou, Ursula Bryan and Daniel Waterhouse, for much practical help and Laura Dekker for assistance with setting up QGAME.

The four node problem definition and QGAME are available via anonymous ftp, site `cs.ucl.ac.uk` directory `genetic/four_node` and a graphical demonstration of the problem can be run via the internet from url
`http://www.cs.bham.ac.uk/~wbl/four_node.demo.html`

References

[Dav91]   Lawrence Davis, editor. *Handbook of Genetic Algorithms*. Van Nostrand Reinhold, New York, 1991.

[FRC93]   Hsiao-Lan Fang, Peter Ross, and Dave Corne. A promising genetic algorithm approach to job-shop scheduling, rescheduling and open-shop scheduling problems. In Stephanie Forrest, editor, *Proceedings of the 5th International Conference on Genetic Algorithms, ICGA-93*, pages 375–382, University of Illinois at Urbana-Champaign, 17-21 July 1993. Morgan Kaufmann.

[FRC94]   Hsiao-Lan Fang, Peter Ross, and Dave Corne. A promising hybrid GA/heuristic approach for open-shop scheduling problems. In A. Cohn, editor, *ECAI 94 Proceedings of the 11th European Conference on Artificial Intelligence*, pages 590–594, Amsterdam, The Netherlands, August 8-12 1994. John Wiley & Sons, Ltd.

[Gol89]   David E. Goldberg. *Genetic Algorithms in Search Optimization and Machine Learning*. Addison-Wesley, 1989.

[Gor95]   T. G. W. Gordon. Schedule optimisation using genetic algorithms. Master's thesis, University College, London, October 1995.

[Hol92]   John H. Holland. *Adaptation in Natural and Artificial Systems: An Introductory Analysis with Applications to Biology, Control and Artificial Intelligence*. MIT Press, 1992. First Published by University of Michigan Press 1975.

[Lan95]   W. B. Langdon. Scheduling planned maintenance of the national grid. In Terence C. Fogarty, editor, *Evolutionary Computing*, number 993 in Lecture Notes in Computer Science, pages 132–153. Springer-Verlag, 1995.

[Lev94]   David Levine. *A Parallel Genetic Algorithm for the Set Partitioning Problem*. PhD thesis, Illinois Institute of Technology, Mathematics and Computer Science Division, Argonne National Laboratory, 9700 South Cass Avenue, Argonne, IL 60439, USA, May 1994.

[LT97]   W. B. Langdon and P. C. Treleaven. Scheduling maintenance of electrical power transmission networks using genetic programming. In Kevin Warwick, Arthur Ekwue, and Raj Aggarwal, editors, *Artificial Intelligence Techniques in Power Systems*, chapter 10, pages 220–237. IEE, 1997.

[PTVF92] William H. Press, Saul A. Teukolsky, William T. Vetterling, and Brian P. Flannery. *Numerical Recipes in C*. Cambridge University Press, Cambridge, 2 edition, 1992.

[Ros94]   Peter Ross. *About PGA 2.8*, 1994. Available via ftp ftp.dai.ed.ac.uk directory pub/pga-2.8.

[RT94]   J. L. Ribeiro Filho and P. Treleaven. GAME: A framework for programming genetic algorithms applications. In *Proceedings of the First IEEE Conference on Evolutionary Computing – Proceedings of the 1994 IEEE World Congress on Computational Intelligence*, volume 2, pages 840–845, Orlando, USA, 26 June-2 July 1994. IEEE Press.

[Sys91]   Gilbert Syswerda. Schedule optimization using genetic algorithms. In Lawrence Davis, editor, *Handbook of Genetic Algorithms*, pages 332–349. Van Nostrand Reinhold, New York, 1991.

[YN92]   Takeshi Yamada and Ryohei Nakano. A genetic algorithm applicable to large-scale job-shop problems. In R Manner and B Manderick, editors, *Parallel Problem Solving from Nature 2*, pages 281–290, Brussels, Belgium, 1992. Elsevier Science.

Solving Generic Scheduling Problems with a Distributed Genetic Algorithm

M. McIlhagga

School of Cognitive and Computing Sciences
University of Sussex, Falmer
Brighton, BN1 9QH.

Abstract. This paper describes a Distributed Genetic Algorithm (DGA) which has been used to solve generic scheduling problems. The GA based scheduler allows the user to define and solve any scheduling problem. It does this using a Scheduling Description Language (SDL). The sort of problem that it might tackle are: job-shop scheduling (JJS), time-tabling, resource sequencing etc. We describe a unique chromosome coding scheme that allows simple representation, straightforward chromosome recombination and fast schedule building and therefore evaluation times. A comparative study has been made of the DGA, random search and a heuristic method of scheduling using 100 very large scale problems; problems of the order of 500 tasks. This is the first study of its kind to look at problems of this scale. It was found that, although it is possible to reduce the makespan of a schedule by about $\frac{2}{5}$ of a randomly generated solution using dispatching rules, only the DGA produced solutions that had as high as a $\frac{3}{5}$ reduction.

1 Introduction

Scheduling has been used by the GA community for a number of years to investigate the application of GAs to a challenging and important class of combinatorial optimisation problems. These studies have concentrated on specific scheduling problems or specific classes of scheduling problems [2]; [10]; [7]. Other recent work, still concentrating on the JSS or similar class problems, have focused on various aspects to improve the performance of GAs in this domain. Yamada and Nakano have unified a multi-step crossover technique with local search to improve the cost of solutions and the effort required to find them [11]. Again Bierwirth *et al* [1] concentrated on JSS to investigate a new permutation representation for the problem. Their representation and crossover technique (precedence preserving crossover) outperforms other techniques, such as PMX. This paper is unrelated to these aspects of GA/scheduling research. The problems that we have been looking at are more generalised large resource allocation problems. The use of more advanced order based representations and crossover operators have not been used in this work. Our primary aim has been to investigate the application of GAs to a wide range of very large non-specific scheduling problems. This paper describes the application of a Distributed GA to *the generic scheduling* problem or the entire class of scheduling problems. We have achieved this by designing and implementing a framework for defining, simulating and solving scheduling problems in a very generalised way.

The results reported here are based on a preliminary testing of this system using 100 diverse large scale problems comparing three scheduling techniques: random search, dispatching rules (a heuristic technique) and a DGA. The problems were generated to reflect the underlying form of JSS that we have tackled previously [7], however they were scaled up to have approximately 50–100 times more schedulable tasks. We found that the random search and dispatching rules methods were able to reduce the makespan of a schedule (using the mean of 10 random solutions as a base for comparison) by about 40 percent. Whereas, the GA was, on average, able to reduce the makespan by 60 percent.

2   A Generic Scheduling System

MOGS (a Model for Optimisation of Generic Schedules) is designed [6] developed to find near optimal schedules for any definable scheduling problem. We have implemented a large part of that model as a 'proof of principal' system; this we call SMOGS (Small MOGS). SMOGS implements all of the concepts in MOGS, but is limited in some functionality. Future work will include the full implementation of the system, eradicating certain inefficiencies in the system and implementing further system attributes to allow the solution of larger and more complex problems.

Section 2.1 details how we used a Scheduling Description Language (SDL) to specify the problems that we are interested in. Section 2.2 outlines how the system simulates the scheduling environment that the problem description (formulated in SDL) represents.

2.1   Problem Description

MOGS incorporates an SDL that enables the user to describe scheduling problems using a set of general concepts. The user's front-end is currently limited: it reads a problem description file containing a scheduling scenario using SDL. Future developments will investigate possibilities of *visual problem description*. Users may find it an easier and more powerful metaphor to use a visual programming tool to formulate problems.

SDL is a relatively simple declarative language (we plan to expand it giving the user more flexibility by including a macro language). It enables the user to describe their problem in terms of *tasks* and their *attributes, resources* and their attributes, *materials* and their attributes, material *flow rates*, material *release rates*, resource and material *location*, etc.

These terms are largely taken from the language of JSS, however there are some significant differences: in MOGS these terms have an exact meaning. In JSS language the term *resource* is badly defined. It usually refers to materials supplied to machines or tools available to machines. Sometimes, however, people or even machines are seen as resources. The term machine is not used at all by MOGS (it uses the term *resource*) and the term *task* largely replaces the terms *job, batch* and *operation* (i.e. unit of work).

MOGS regarded each scheduling problem as a set of tasks that need to be completed. Resources are those things that complete tasks (machines, teams of people, individuals etc.). Each task has a varying number of attributes, e.g.: when it must be completed by, the required attributes of a resource capable of completing that task etc. Resources also have attributes: type of task attributes that they process etc. These attributes might be a machine operation such as drilling (JSS) or they might be a skill, such as experience with SQL data-base languages (human resource scheduling) or a rooms seating capacity (room timetabling).

MOGS also models (or schedules) the supply and processing of *materials* to and by resources in order to compete tasks. In Job-Shop terms this includes the supply of raw materials to machines, for example plastics or metal blanks. It also includes the supply and fitting of tool parts (these are regarded by MOGS as a special case of material supply, i.e. *reusable* materials). Materials are supplied from reservoirs; represented as queues or stacks. These are filled according to data suppled by the user which reflects the actual supply of those materials, e.g. as deliveries from third parties or as supplied by the output of some resource. As the completion of a task is the result of a resource *acting* on some materials to *produce* another material, it is possible to set up a hierarchical supply and demand situation between a number of resources. This allows the user to implicitly define ordering constraints (they can also do this explicitly if they wish). For instance, an assembly situation can be described by simply stating the input/output relationship between various tasks and from these SMOGS will derive those necessary constraints in the problem.

An example of this might be the assembly of a paper stapler from as in figure 1. We can define an initial task of making the *stapling mechanism* as assembling the main body of the stapler to plate that slides inside it and attaching a spring to form the return mechanism. This is attached by a hinge and another spring to the bottom of the stapler to form the finished object. This sort hierarchical process of manufacture is at some level common to the production of most of every-day objects in our lives. Because MOGS models the use of *materials*, simply defining the tasks also defines many of the necessary ordering constraints in the problem.

This aspect of the system adds to the dimensionality of the problem. Although the problems we have used for this study are based on job-shop problems they also include the aspect of material supply: transport rate, stock control etc. This more accurately models the real scheduling problems that industry encounters. These problem are in the class of JSS problems, but are much harder than those usually tackled [9] because they have been reformulated to include material scheduling.

One thing that the user cannot do is tailor the internal representation of problem to suit a specific problem. It is unclear whether this is a disadvantage or not. One could take the view that specifically designed representations will

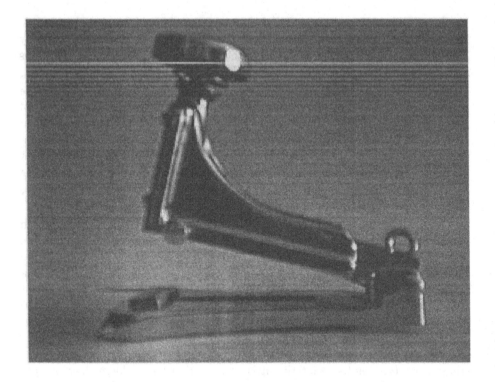

Fig. 1. Assemblage of a Paper Stapler

always outperform the general case, however if all scheduling problems share attributes then a generalised algorithm may not perform *significantly* worse than specialised solutions. It cannot be pragmatic to research a new representation and parameter set for each new problem where a general solution is both more immediate and not significantly worse. However, this may not be the case. Future research should address these issues as they are directly pertinent to the needs of industry along with issues such as solution robustness and scheduling in reactive environments.

2.2 Objective Functions

SMOGS reads the problem description and creates a model of the scheduling environment. Candidate solutions are created by one of the search techniques available, presently: random search, dispatching rules or a DGA. One (or a combination) of a number of possible objective functions are used to determine the worthiness of that schedule. The objective function is a *discrete event simulation* which builds schedules by decoding chromosomes evolved by the DGA. These are mapped into a 'gantt chart' (cf. [3]) like data structure via a 'resource availability graph' (see section 2.3). The resource availability graph is built only once at the beginning of each run. At the present time the user can set the objective function to any one or user defined combination of: makespan, mean flow time,

release date	RD_j
a due date	DD_j
a completion time	CT_j
flow time	$FT_j = CT_j - RD_j$
lateness	$LA_j = CT_j - DD_j$
tardiness	$TN_j = \max(0, LA_j)$.
Duration of TA_j on RE_i	Du_{ij}
makespan or total schedule time	$MS = CT_{max} - RD_{min}$
mean flow time	$MFT = \mathrm{sum}(j=i, j=n, FT_j/n)$
total tardiness	$TTN = \mathrm{sum}(j=1, j=n, TN_j)$
total lateness	$TLA = \mathrm{sum}(j=1, j=n, LA_j)$
initial resource availability date	AD_i
resource utilisation	$RUi = \mathrm{sum}(j=1, j=n, DU_{ij}/(MS-AD_i))$

Table 1. Cost Functions

resource utilisation, proportion tardy, total tardy, total lateness, total earliness, mean earliness, mean 'on-timeness'. Other less JSS like objectives will be made available in later versions.

Total earliness is the sum of the amount of time each task is completed early (lateness not included). Mean earliness is total earliness divided by the number of tasks. Mean on-timeness is the mean (modulus) difference between the expected completion of tasks and their actual completion. All other terms are the more usual metrics found in the literature [3], [8], and are defined in table 1.

Given n tasks, $TA_j (1 \leq j \leq n)$. Where the tasks are to be carried out by r resources, RE_i, $(1 \leq i \leq r)$. For a given schedule TA_j there exists a number of useful features and statistics for that schedule, see table 1.

2.3 Resource Availability Graph

When the problem description is read from a file, a data structure know as the resource (availability) graph is created and modified according to the problem at hand. The resource graph is essentially a decision tree. Possible paths through which are defined by each legal chromosome. The tree limits the space of all possible solutions to the space of legal solutions: matching viable resources to each task and possible supplies of material to each resource. Each branch of the tree has a 'logical' index which is used to map chromosome genes to 'physical' resources, reservoirs and materials etc. The upshot of using this scheme is that all genes in all chromosomes are inevitably legal and the only search space that can be represented is a legal one. Moreover, aspects of the encoding (see section 4) allow random bit mutation and simple crossover to be applied without fear of corrupting the legality of the chromosome. This meant that a GA tool kit could be used to implement the search aspect of the system. Only the crossover function that combines the task orders had to be written specifically for this application.

3 Search Techniques

3.1 Random Search

The Random search implemented in this study works by finding 30,000 consecutive random solutions to each scheduling problem and storing the best as they are generated. To generate a random solution it was necessary to call the random number generator some 1500 times per schedule.

3.2 Dispatching Rules

Dispatching or priority rules are a popular heuristic used in constructing schedules in many types of scheduling problems [3]. Often used within simple constructive search algorithms to find the next operation to process, the most common are: SPT, Shortest Processing Time; FCFS, First Come First Served; MWKR, Most Work Remaining; LWKR, Least Work Remaining; MOPNR, Most Operations Remaining; RDM, Random.

As a heuristic technique for building schedules we currently use of a dispatching rule based on slack time (the difference between time remaining to due date and anticipated total process time). Whenever a resource becomes available, the task chosen to be processed next is the one with least slack time (LST). Slack time is a concept that applies to a task. In order to decide between candidate task/resource combinations, we make use of the SI (shortest imminent processing time) rule. Thus the general scheduling problem is tackled by using a LST rule to choose between tasks, and the SI rule to choose the task-resource combinations. This technique is one with which we have made comparisons before. It was first used by Khoshnevis and Chen [4] and later by Palmer [8]. Recently we compared Palmer's comparison of Khoshnevis and Chen's Dispatch Rules and Simulated Annealing with a distributed GA approach [7]. Further heuristics will be experimented with in later versions of the system, but these rules are applicable to a very wide range of problem types.

3.3 Genetic Algorithm

The Genetic algorithm uses a fairly standard distributed model where each chromosome has a 2D location on a toroidal grid (see figure 2).

The GA was parameterized in a similar way to previous work [7]. However, due to the relatively long evaluation time ($\approx$1 second per schedule) the population was reduced, from a more ideal size of 1600 chromosomes, to 300 chromosomes. Other parameters were:

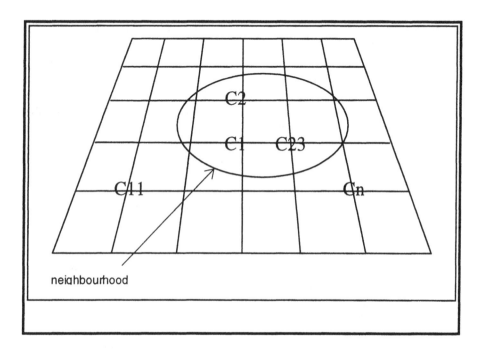

Fig. 2. A Distributed GA Uses a 2D Geographical Grid for Breeding

- Parent selection: ranked local section with a neighbourhood of the 12 nearest. The nearest 12 chromosomes to an initial randomly selected parent are placed in a selection pool. The pool is ranked in fitness order and a standard rank selection is applied to choose the second parent.
- Replacement strategy: replace worst parent. There is only a finite amount of memory allocated to the population. Each allocated slot for a chromosome is always in use. Creating a new chromosome means deleting one already in the population. The memory that the worst parent of the new chromosome occupies is over-written by its new offspring, however its geographical location may not be the same as its now deleted parents.
- Placement strategy: place new chromosomes 'random-local' to their parents. An empty random location is selected within the neighbourhood used to choose the second parent of the new chromosome. This becomes the new location of the child chromosome.

- Operators: crossover (see section 5) and mutation at a rate of 0.056 (this is the probability that **any** one bit in each 'gene' is flipped).

We used GPDGA[1], the Generic Parallel Distributed Genetic Algorithm tool kit to implement the evolutionary search aspect of this study [5].

4 Genetic Encoding

The chromosome is coded as an array of integers, as in table 2. It is split into three sections. First, the ordering of tasks is represented. Second, the resources to be allocated to each task are encoded. Finally, the reservoirs that will supply each of the materials required by the resource for each task are encoded. The chromosome is of a fixed length for any one problem, but varies between problems. The chromosomes is of a fixed length because there are a defined number of tasks in a problem and only one resource will be allocated to each tasks (i.e $2 \times T$ genes). However, the number of materials need for each task can vary, so it is not possible to determine the chromosome length from just a cursory look at the problem description. The third section maps an appropriate material supply reservoir for *each* material needed to complete each task.

If we say that T is the number of tasks in the problem and that each integer is a gene then (refer to table 2):

1. The first T genes contain integers in the range $0 < Tn < T$ denoting the ordering of tasks, i.e. how they are place on the gantt chart. The ordering not only defines the precedence of one task over another on a given resource, but also assigns precedence of material supply to tasks earlier in the order, (even though their time slot might be later than a task later in the order which is placed on a less utilised resource).
2. The next T genes in the chromosome denote the resources that tasks will use to complete. The format is such that the first resource is allocated to task 1 in the problem data structure, not the first task in the ordering defined in the first part of the chromosome.
3. The rest of the chromosome is dedicated to mapping which reservoir will supply what material to which task–reservoir pair. A reservoir can be thought of as a buffer that releases materials to (in the case of a job-shop, machines) the resources which process tasks. It is mapped out in the following way:

for each task in the problem data structure
    and for each material required by that task
        there is a gene that denotes the reservoir
        that will supply the material.

[1] GPDGA was developed under the EPSRC funded project no GR/J 40812.

Example 1: This very simple problem has 8 tasks, 1 of which requires 3 materials, 2 of which requires 2 materials, the rest requiring only 1. The chromosome structure would look like table 2.

CHROMOSOME STRUCTURE																							
format	G	G	G	G	G	G	G	G	G	G	G	G	G	G	G	G	G	GG	GG	G	G	G	GGG	G
type	O	O	O	O	O	O	O	O	T	T	T	T	T	T	T	T	T	TT	TT	T	T	T	TTT	T
map	N	N	N	N	N	N	N	N																
label	task order								resources								reservoirs							
number	8								8								12							
bytes	32								32								48							

Table 2. Example Encoding of a Chromosome.

4.1 Encoding Types

As table 2 suggests, the encoding for the first part of the chromosome is different from that of the second and third part. The O (table 2) or Order Type of encoding contains a number in the range 1 to the number of tasks (T) and the whole section contains 1 (and no more) of each number in that range. This encoding requires special crossover and mutation routines (see section 5). This section of the chromosomes represents the order and therefore the priority with which tasks are placed into the schedule. This process of constructing the schedule should not be confused with the actual scheduled times of each task. It is quite possible for the chromosome to place the task that is scheduled to start last onto the schedule data structure first; it simply monopolises its resource and, as a result, will not have to compromise the start and completion dates that it wanted.

The numbers represented in the second and third section of the chromosome do not map to resource and material indices described by the user in SDL. Rather, they map to legal resources and reservoirs for each task in the *resource graph*, making each gene by definition legal. All bits are set in these genes as a many to one representation is used (this needs decoding to determine the encoded resource).

The T or Toroidal type was especially formulated to allow the representation of varying ranges of numbers using a blind encoding. That is, all T types can have any bit in the gene set and still maintain a legal encoding of any range of integer values (with the proviso that numbers are within the combinatorial range of the integer on that specific machine). This is achieved by making all binary values that the gene can hold legal and by taking the integer remainder of a division of the chromosome by the maximum value of its range (T) as the decoded value. This causes a minor problem for mutation, but one that is simply solved (see section 5). In order to implement this model a chromosome map or template must be maintained that describes the range of numbers in each gene (see Table 2). In this case the range of e.g. resources, denotes the list of viable resources for that task detailed in the resource availability graph data structure and **not** the list of all resources in the problem. Using this encoding with a graph of this type in the evaluation module forces any gene value to map to a resource that has the right attributes to enable it to carry out that task; illegal encodings are therefore not possible.

The upshot of using these representations and encoding is that the process of breeding and evaluation is very simple to implement and fast to execute. Much of the schedule building work has been 'taken out of the loop' by building the resource availability graph and the chromosome 'range' map at the start of the run. Thus enabling the solution of larger problems. The real significance of this rather specific encoding is that it is not specific to any sub-class of *the generic scheduling problem*, but encodes for the whole class of problems.

5   Operators

Crossover in the first (type O) section of the chromosome works in the following way: a sub-string of one parent is found and inserted into the other parent once the items in the sub-string have been removed from the receiving parent. Substring length, position and insertion point are chosen at random.

Crossover for type T is more straight forward. It is implemented in the following way. N crossover points are chosen in the parent chromosomes. Often this will be one crossover point per chromosome, however this can be defined by the user. The section before the crossover point in parent A is concatenated with the section after the crossover point in parent B to form the new child chromosome.

Mutating the first section of the chromosome, which represent a unique ordering of tasks, is more problematic than bit mutation of the binary strings used for the second and third section of the chromosome. Mutation of an ordered set can take a number of forms. In all cases, the restriction that one of each of the numbers in the range $\{1, T\}$, where T is the number of genes in the ordered section of the chromosome, must hold. This can be achieved by implementing 1 or all of:

1. Swapping the order of two juxtaposed tasks in the chromosome.
2. Swapping the order of two task allocated the same resource.
3. Moving a higher priority task on Resource N to just after the juxtapose (lower priority) task on Resource N.

Mutation of type T genes can be done in two ways:

1. Mutate one or more bits. This can cause big or small changes in what any give gene represents and there is no way of knowing what that might be by.

having the bit position in the gene. This is not a problem as this method is functionally equivalent to random mutation in the working bit range of a chromosome that is directly encoded.

2. When using a direct representation, random walk or Lower order mutation (near the LSB of gene) is equivalent to (e.g. for lower three bits) the addition of one of the following set of numbers chosen at random : $\{1, -1, 2, -2, -4, -4\}$. This sort of mutation can be implemented by decoding the chromosome and making the random addition and re-encoding the new value, taking into account the result of the integer division of the chromosome by the range (see section 4).

6 Results of Study

6.1 Description of Study

A comparison of random search, dispatch rules and a distributed GA as techniques for the solution of large scale general scheduling problems was made. The set of problems described below are modelled on standard JSS problems. However, they are much more than that. Because of the way MOGS models material flow a more natural representation of ordering constraints is present. Another difference between these and many JSS problems is that resources are not already allocated to tasks. These factors combined with the scale of the problems in terms of the number of tasks means that we are dealing with much harder problems than 'tradition' JSS.

The problem set that we used comprised 100 scheduling problems, i.e. 100 files containing a lengthly description of a problem using our SDL. Each problem had between 400 and 500 tasks (randomly distributed throughout the 100 problems).

Typical run times on a PentumPro-180 were in the order of 2 hours (30,000 evaluations) for the random search and the evolutionary approach and 125 seconds for the heuristic approach. The latter figure is due to the fact that the heuristic approach has two phases. First, the assignment phase, where tasks are assigned to resources. Here the calculation of where to place each (and which) task take roughly the same time as a schedule evaluation: about 0.25 of a second. The second phase is a full evaluation, again taking approximately 0.25 of a second. The time for the construction of a single schedule of 500 tasks is therefore approximately 125 seconds.

With these facts in mind it is easy to assess the use of a heuristic approach as better than that of random search. Under certain time critical circumstances it is also better than an evolutionary approach. This would only be the case where the schedule is required before the more computationally intensive technique (the genetic algorithm) is capable of exceeding the quality of the solution found by heuristic means. The heuristic technique cannot be improved by adding more evaluations: this leaves the comparison of these techniques difficult. Clearly there are situations where it is unacceptable to wait two hours for a schedule, conversely there will also be situations where cost factors dictate the use of a more computationally expensive algorithm that offers better solutions.

Figure 2 below depicts the crossover point at which this would occur. The significant parameters that combine to define the search space of each problem are:

- The number of tasks, the number of resources capable of completing each task (typically 4-10 in this study).
- The number of reservoirs capable of supplying each of those resources (1,2 or 3 in this study).
- The number of attributes that each tasks has (upto 10).
- The number of input materials and material flow as defined by resource location, release rates and travel rate.

A typical search space is 1×10^{1500} solutions. These problems are only limited by the parameters of this study and are by no means the limit of the system. Memory and processor speed allowing, it is possible to define a scheduling problem of any size — there is no practical upper limit on the number of tasks, resources, or materials etc. Because we can easily define a variety of resource allocation or scheduling scenarios, future work will investigate what the limits of problem size is in terms of optimisation. Given a set limit on computational effort, preliminary tests indicate that the costs of very large problems found by GA tend to the values found by heuristic approach detailed above, suggesting that hybrid approaches may be the only way forward for the optimisation, rather than solution, of very large scheduling and allocation problems.

6.2 Results

Table 3 shows all of the figures for percentage **decrease** of makespan (from the mean of ten random solutions) for all techniques: the comparative decrease made by the DGA is 17.178 percent over heuristic search and 20.7547 percent over random search.

MAKESPAN PERCENTAGE DECREASE				
	Mean	Std. Dev.	Min	Max
Random	39.2580	14.0824	22.8734	62.4510
Heuristic	42.8347	12.0045	18.9035	64.9084
Evolutionary	60.0127	13.7439	19.2982	75.9534

Table 3. Statistic for Percentage differences in Makespan (all Techniques)

The evolutionary technique produces decreases in makespan of 20.7547 percent over random search. This figure, as can be seen from Table 3, is a highly significant improvement on the heuristic (dispatch rule) approach. We also made a comparison of relative resource slack time between schedules generated by the different techniques. Slack time is simply the time that the machine is not utilised by a task. This varies slightly from resource utilisation, as a task may hold a resource unutilised as it waits for materials.

We found that the evolutionary approach could make better use of the resource efficiency attribute of each resource, and thereby leave less resources standing idle during the period that the schedule covers. Table 4 shows the differences between each of the techniques.

MEAN PERCENTAGE RESOURCE SLACK TIME	
	Mean
Random	20.8234
Heuristic	18.7123
Evolutionary	12.5137

Table 4. Mean Percentage Slack Time for Each Resource

The reduction of resource slack time and therefore related (but not the same) resource (machine) utilisation is a significant indication of why the DGA was able to find better schedules than the other techniques.

7 Conclusions

Because the mean improvement in makespan using random search (from a mean of 10 randomly generated solutions) is 39 percent, it would be foolish to entirely dismiss this technique.

The mean improvement in makespan using a heuristic scheduler is 42 percent. Although this figure may not appear to be that much better than random search, if the improvement were to be divided into computational time taken to find solutions we find that the heuristic method is exceptionally efficient.

It is the case that where the schedule is required before the more computationally intensive technique (the genetic algorithm) is capable of exceeding the quality of the solution found by heuristic means, that heuristic search will out perform all other techniques. However this time window is very small and in most cases we can take advantage of an evolutionary approach.

Of the techniques investigated, only the Genetic Algorithm can provide useful solutions with little resource slack time. Current literature does not contain any study where the problems are of this order of magnitude. Nor do any other systems use a completely generic approach to scheduling problems.

These problems are extremely difficult to tackle. The search spaces are of the order of 1×10^{1500} solutions where each solution takes up to a second to evaluate. We can only hope to search a very tiny percentage of the search space, so appropriate direction as to where to search is paramount. It is this sort of problem that distributed GAs seem to thrive on. Although the solutions found may be sub-optimal, there is now no doubt that powerful evolutionary search techniques, such as those developed for this system, are capable of producing solutions of a quality that far exceeds that of those produced by the traditional techniques in place in industry today. This problem set has really stretched the limit of what is possible with current GA technology.

Future work will include a full implementation of the MOGS system design. Although this paper has not covered the difference between the MOGS specification and the SMOGS implementation, various classes of scheduling problems can be easily specified using MOGS (which supports a hierarchical task structure) that are difficult to represent and solve using SMOGS. A full implementation of MOGS will also provide a faster and more effective problem modeller.

References

1. C. Bierwirth, D. C. Mattfeld, and H. Kopfer. On permutation representations for scheduling problems. In I. Rechenberg H. Voigt, W. Ebeling and H. Schwefel, editors, *Parallel Problem Solving From Nature - PPSN IV*, volume 1141 of *Lecture Note in Computer Science*, pages 960–970. Springer-Verlag, Berlin, 1996.
2. L. Davis. Job-shop scheduling with genetic algorithms. In J. Grefenstette, editor, *Proc. Int. Conf. on GAs*, pages 136–140. Lawrence Erlbaum, 1985.
3. S. French. *Sequencing and scheduling: an introduction to the mathematics of the job-shop*. Ellis Horwood, 1982.
4. B. Khoshnevis and Q. Chen. Integration of process planning and scheduling functions. *Journal of Intelligent Manufacturing*, 1:165–176, 1990.
5. M. McIlhagga. A generic encoding for scheduling problems. Technical report, University of Sussex, 1995.
6. M. McIlhagga. Gpdga user documentation. Technical report, University of Sussex, 1995.
7. M. McIlhagga, P. Husbands, and R. Ives. A comparison of simulated annealing, dispatching rules and a coevolutionary distributed genetic algorithm as optimization techniques for various integrated manufacturing planning problems. problem. In *Proceedings of PPSN IV*, volume LNCS, 1141, pages 604–613. Springer Verlag, 1996.
8. G. Palmer. *An Integrated Approach to Manufacturing Planning*. PhD thesis, University of Huddersfield, 1994.
9. K. Sycara, S. Roth, and M. Fox. Resource allocation in disributed factory scheduling. *IEEE Expert*, pages 29–40, Feb. 1991.
10. H. Tamaki and Y. Nishikawa. Parallelled genetic algorithm based on a neighborhood model and its application to job-shop scheduling. In *Proceedings of PPSN II*. Springer Verlag, 1992.
11. T. Yamada and R. Nakano. Scheduling by genetic local search with multi-step crossover. In I. Rechenberg H. Voigt, W. Ebeling and H. Schwefel, editors, *Parallel Problem Solving From Nature - PPSN IV*, volume 1141 of *Lecture Note in Computer Science*, pages 960–970. Springer-Verlag, Berlin, "1996".

Directing the Search of Evolutionary and Neighbourhood-Search Optimisers for the Flowshop Sequencing Problem with an Idle-Time Heuristic

Peter Ross and Andrew Tuson

Department of Artificial Intelligence, University of Edinburgh
80 South Bridge, Edinburgh EH1 1HN, U.K.
{peter,andrewt}@dai.ed.ac.uk

Abstract. This paper presents a heuristic for directing the neighbourhood (mutation operator) of stochastic optimisers, such as evolutionary algorithms, so to improve performance for the flowshop sequencing problem. Based on idle time, the heuristic works on the assumption that jobs that have to wait a relatively long time between machines are in an unsuitable position in the schedule and should be moved. The results presented here show that the heuristic improves performance, especially for problems with a large number of jobs. In addition the effectiveness of the heuristic and search in general was found to depend upon the neighbourhood structure in a consistent fashion across optimisers.

1 Introduction

One problem of interest in the artificial intelligence and operations research scheduling communities is the flowshop sequencing problem. The aim of this problem is, quite simply, to find a sequence of n jobs that minimises the *makespan* — the time for all of the jobs to be processed by m machines. This paper will examine two ways how domain-knowledge can be included into evolutionary and stochastic neighbourhood search techniques: in the neighbourhood structure, and in biasing the order in which different possible moves (mutations) are tried to examine whether such knowledge is transferable between techniques; with an emphasis upon the second of these aspects.

The rationale for the second aspect is that, for each solution in the search space, there are number of 'moves' to other solutions available, most of which are non-improving. Conventional implementations examine these moves in a random order; however; if a method existed that could cheaply determine which moves were improving, it would then result in high-quality solutions being found more quickly (assuming that the search would not be led into a local optimum).

In practice, such a method would be heuristic in nature. Fortunately, it is often possible to examine a solution and to form beliefs about which moves would be most likely to improve the solution. These beliefs can then be captured in a heuristic, which makes it more likely that such moves are attempted first. The approach used here was based on the 'directed mutation' operator [15] that has been successfully used in timetabling, where exams with a high number of constraint violations were more likely to be moved; it is also related to the work by [18,19] on iterative repair scheduling, and by [9] on the 'min-conflicts' heuristic.

For this problem, such a belief can be formulated by using the idle times, the time a job spends waiting between machines, associated with each job in the sequence to decide what the most profitable moves are likely to be. Therefore, the effectiveness (or otherwise) of using the idle times as a heuristic to direct the search of some stochastic optimisation techniques will be examined. It should be noted that the idea of using idle times to improve the performance of heuristic methods for the flowshop sequencing problem is not new: [5] proposed the use of a similar heuristic in an approach that is essentially a deterministic hillclimber, and [13] used idle times in a constructive algorithm. Also, the concept of a 'candidate list', which orders the sequence in which the moves are considered has been used by [11], and in the evolutionary computing literature by [17].

However, the authors are unaware of an idle-time heuristic being used for evolutionary and stochastic neighbourhood search optimisation approaches, in this way, for this problem and whether, or to what degree, experience of representing domain knowledge gained with one optimisation technique can be transferred to another. Thus these will be main topics under investigation in this paper. The problem and the optimisers that are the subject of this investigation will be outlined first. The heuristic under investigation and how it is used to direct the search will then be described, and then the results obtained evaluated across a range of optimisers, and finally conclusions drawn.

2 The Flowshop Sequencing Problem

The flowshop sequencing, or $n/m/P/C_{max}$, problem involves finding a sequence of jobs for the flowshop (a straight line of machines to process, so as to minimise the *makespan* — the time taken for the last of the jobs to be completed. The makespan is dependent upon the order in which the products are produced because poor sequences can lead to temporary blockages in the flowline because completed stages cannot proceed, as a stage ahead in the flowline is still being processed (Figure 1).

This task is known to be NP-hard [7] (the number of possible sequences is $n!$) and can be formalised as follows: n jobs have to be processed (in the same order) on m machines; the aim is to find a job permutation $\{J_1, J_2, ..., J_n\}$ so as to minimise C_{max}. This is defined as follows: given processing times $p(i, j)$ for job i on machine j and the job permutation above, we can find the completion times by the following equations:

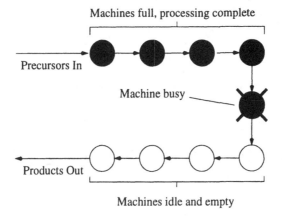

Fig. 1. Blocking of a Flowshop at a Single Machine

$$C(J_1, 1) = p(J_1, 1)$$
$$C(J_i, 1) = C(J_{i-1}, 1) + p(J_i, 1) \; for \; i = 2, ..., n$$
$$C(J_1, j) = C(J_1, j - 1) + p(J_1, j) \; for \; j = 2, ..., m$$
$$C(J_i, j) = max\{C(J_{i-1}, j), C(J_i, j - 1)\} + p(J_i, j) \; for \; i = 2, ..., n; \; j = 2, ..., m$$
$$C_{max} = C(J_n, m)$$

Standard benchmarks exist for this problem, and were used in this study. Taillard [16] produced a set of test problems which, using a very lengthy Tabu Search procedure, were still inferior to their lower bounds, with problem sizes ranging from 20 jobs and 5 machines, to 500 jobs and 20 machines. There are 10 instances of each problem — all processing times were generated randomly from a $U(1, 100)$ distribution.

3 The Optimisation Techniques Considered

A variety of stochastic optimisation methods (an overview is given in [4]) were examined: Stochastic Hillclimbing (SHC), Simulated Annealing (SA) [8], Threshold Accepting (TA) [3], Record-to-Record Travel (RTRT) [2], and an Evolutionary Algorithm (EA) [6]. Basic implementations were used in this study as the aim was to investigate whether the neighbourhood could be usefully directed, and not to find the most effective optimiser.

The encoding used for this problem was a permutation of the jobs to be placed into the flowshop, numbered 1 to n. Two *unary* neighbourhood (ie. mutation) operators were used in this study: permutation-shift, and permutation-swap (Figures 2 and 3), where the jobs that are bounded by a rectangle are the two jobs selected to be moved by the operator.

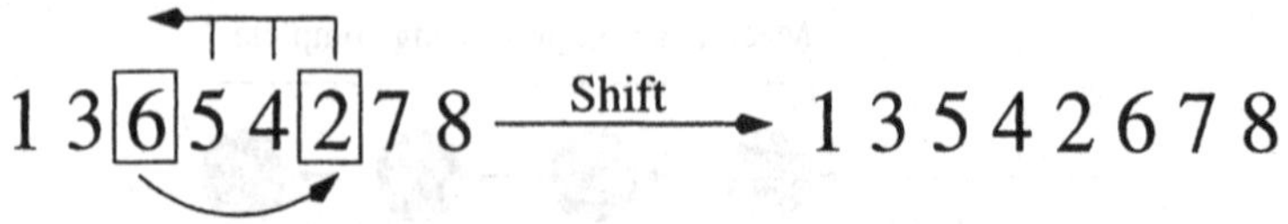

1 3 6 5 4 2 7 8 —— Shift —→ 1 3 5 4 2 6 7 8

Fig. 2. The Permutation-Shift Operator

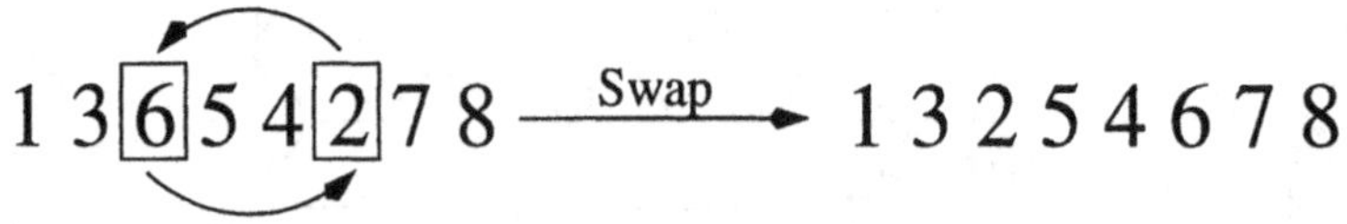

1 3 6 5 4 2 7 8 —— Swap —→ 1 3 2 5 4 6 7 8

Fig. 3. The Permutation-Swap Operator

It should be noted that although the operation $swap(i,j)$ is equivalent to $swap(j,i)$, the same is not the case for $shift(i,j)$. In the case of the $shift(i,j)$ operator, position i (which is job number 6 at position 3 in the example above) corresponds to the job that is displaced to make room for the block of jobs that is to be shifted, the extent of which is given by position j. Exchanging the values of i and j in a shift operation therefore leads to a reversal of the roles of the two positions, and of the direction that the block is shifted.

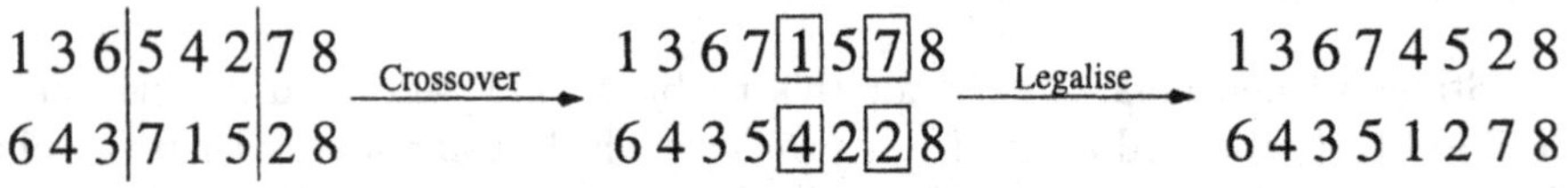

1 3 6 5 4 2 7 8 —— Crossover —— 1 3 6 7 1 5 7 8 —— Legalise —— 1 3 6 7 4 5 2 8
6 4 3 7 1 5 2 8   6 4 3 5 4 2 2 8   6 4 3 5 1 2 7 8

Fig. 4. The Modified-PMX Crossover Operator

The crossover operator used for the EA was 'Modified PMX' [10], found to give good results on this problem. This operator performs a two-point crossover upon the two strings selected. The repair procedure then analyses one string for duplicates: when one is found it is replaced by the first duplicate in the second string. This is repeated until both strings are legal (Figure 4).

4 The Idle Time Heuristic

The heuristic used here is based upon idle times. Idle time is usually defined for a particular job and machine, and is the amount of time that a job is waiting to be processed at a machine after it has been processed by the previous machine. This is illustrated by the Gantt chart (Figure 5) below — the gaps between blocks are the periods which a job is lying idle. However, as we are making a decision about which job to move, it would be sensible to add up the idle time for each job, over all machines, as given in the equations below:

$$I(J_1) = \sum_{j=2}^{m} p(J_1, j-1) \ (for \ the \ purposes \ of \ this \ study \ - \ see \ below)$$

$$I(J_i) = \sum_{j=2}^{m} max\{C(J_{i-1}, j) - C(J_i, j-1), 0\} \ for \ j = 2, ..., m$$

with processing times $p(i,j)$ for job i on machine j, the job permutation $\{J_1, J_2, ..., J_n\}$, completion time $C(J_i, j)$ and idle time $I(J_1)$.

Note that the first job in the sequence is given an idle time equal to the sum of the processing times on the machines $2, ..., n$ for that job. This is because the idle time for the first job in the sequence is not defined as the machines in front of it are empty. Fixing this to zero would have the effect of ensuring that the first job always has a low probability of being moved. As it is unlikely that the first job in the initial solution's sequence will be the correct choice, and so should be moved eventually, it is assigned the amount of idle time given above.

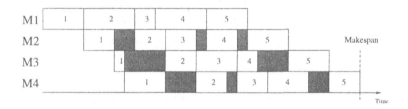

Fig. 5. An Example Gantt Chart for a 5x4 Flowshop

How can this information be used to predict which moves are most likely to lead to improvements in makespan? The answer is that first, having jobs waiting to be processed is not desirable as they represent delays; second, it is reasonable to suppose that jobs that are left waiting for a long time are in an unsuitable position in the sequence, and therefore should be moved.

The evaluation function was adapted to return the idle times associated with each job, as described above. When the neighbourhood operator was then applied, the position(s) of the move operator was decided on the basis of the idle time of the job at that position, by using one of either Tournament Selection [1] or Marriage Selection. Marriage Selection is a variant of tournament selection designed to soften its inherently high selection pressure. Possible jobs are picked at random until either one is found that is more attractive than the first, or a fixed number (the *Monte Carlo* size) of possible moves have been selected, in which case the job with the highest idle time associated with it is chosen [14].

5 Evaluating The Heuristic

The measure of performance used was the quality of solution obtained after a set number, N, of evaluations. Fifty runs were taken in each case, therefore summaries are given below and detailed results are available from the authors. Experiments were performed for the first of the Talliard test set's [16] instances of the following ($n \times m$) problems: 20x5, 20x10, 20x20, 50x5, 50x10, 50x20, 100x5, 100x10, 100x20, 200x10, and 200x20. The value of N used was solely dependent upon n and was set to 5000, 6750, 8000, and 9250 evaluations respectively. This is a roughly $ln(n)$ relationship, justified from empirical results from [12].

5.1 Stochastic Hillclimbing Results

Problem	Undirected	Heuristic + Random	Heuristic + Heuristic
20x5	1296.26 (3.63)	**1291.68 (7.52)** [T9]	**1294.58 (5.68)** [M4]
20x10	1633.32 (21.73)	1629.32 (21.15) [M5]	*1642.66 (27.37)* [T2]
20x20	2373.16 (26.94)	2367.80 (29.29) [M3]	2372.52 (23.68) [T2]
50x5	2742.82 (9.14)	**2728.90 (4.10)** [T9]	**2732.56 (6.62)** [T4]
50x10	3141.72 (24.76)	**3126.72 (19.98)** [M8]	3133.68 (25.85) [M3]
50x20	4049.34 (31.64)	**4037.62 (26.40)** [T3]	4046.38 (25.83) [T2]
100x5	5510.62 (12.91)	**5500.08 (11.75)** [T8]	**5495.80 (4.28)** [T8]
100x10	5927.38 (37.54)	**5893.12 (35.40)** [T9]	**5907.42 (33.17)** [M6]
100x20	6622.44 (44.06)	**6588.62 (35.34)** [M6]	**6605.48 (32.09)** [M4]
200x10	11029.90 (35.02)	**10995.12 (33.04)** [T9]	**10992.40 (28.64)** [M8]
200x20	11826.10 (49.64)	**11764.14 (59.65)** [T8]	**11780.46 (54.43)** [M6]

Table 1. Stochastic Hillclimbing Results — Swap Neighbourhood

The results for stochastic hillclimbing are given in Tables 1 and 2. For all of the tables presented here, the mean and standard deviations of the makespan are given, with the selection type and pressure found to be most effective in square brackets.

Problem	Undirected	Heuristic + Random	Random + Heuristic	Heuristic + Heuristic
20x5	1293.60 (7.26)	1292.58 (7.81) [M4]	**1284.00 (10.32)** [T9]	1288.52 (9.27) [M3]
20x10	1618.92 (13.13)	1615.80 (10.62) [T2]	1615.12 (16.12) [T2]	1617.34 (16.64) [T2]
20x20	2358.88 (33.10)	2361.28 (25.90) [M3]	2352.30 (26.25) [T2]	2364.22 (27.15) [T2]
50x5	2734.46 (6.76)	**2731.98 (6.93)** [T8]	2730.48 (5.15) [M5]	**2730.92 (6.47)** [T4]
50x10	3121.10 (24.73)	3125.86 (19.89) [M8]	**3107.96 (21.59)** [T3]	3117.54 (18.77) [M3]
50x20	4021.62 (22.52)	4019.64 (23.58) [M5]	**4011.98 (25.19)** [M3]	4018.68 (21.74) [T2]
100x5	5506.56 (14.14)	**5498.70 (8.53)** [M5]	**5493.24 (0.65)** [T9]	**5493.98 (1.50)** [T8]
100x10	5884.80 (30.06)	*5895.82 (35.19)* [T2]	**5842.00 (20.49)** [T8]	5874.08 (32.51) [M6]
100x20	6581.78 (38.13)	6582.32 (36.89) [M6]	**6542.88 (36.09)** [T5]	**6556.70 (38.44)** [T2]
200x10	11020.20 (31.61)	**11007.94 (35.52)** [M5]	**10977.52 (38.20)** [T9]	**10982.40 (23.85)** [T6]
200x20	11797.80 (50.06)	**11770.14 (51.45)** [M3]	**11669.52 (48.67)** [T8]	**11715.88 (53.82)** [M5]

Table 2. Stochastic Hillclimbing Results — Shift Neighbourhood

The form of the information about the selection method is as follows: the capital letter represents the selection method used (M for marriage, T for tour-

Problem	Undirected	Heuristic + Random	Heuristic + Heuristic
20x5	1296.00 (4.30)	**1293.84 (6.54)** [T6]	1294.98 (5.29) [T3]
20x10	1619.72 (14.73)	1619.52 (17.07) [T6]	*1626.76 (15.30)* [T2]
20x20	2357.36 (24.21)	2355.22 (22.79) [T2]	2359.44 (21.21) [T2]
50x5	2741.14 (10.02)	**2730.84 (6.04)** [T9]	**2735.64 (7.14)** [T4]
50x10	3135.34 (21.37)	**3122.84 (22.93)** [T4]	3131.44 (20.66) [T2]
50x20	4044.52 (27.69)	**4027.76 (24.04)** [T5]	**4030.72 (25.49)** [T2]
100x5	5516.12 (13.28)	**5500.32 (10.63)** [T9]	**5498.40 (8.56)** [T9]
100x10	5939.50 (41.34)	**5880.32 (24.12)** [T8]	**5902.92 (24.74)** [M4]
100x20	6606.50 (31.98)	**6582.76 (31.96)** [T4]	**6586.84 (28.20)** [T2]
200x10	11035.90 (41.03)	**10992.02 (26.15)** [T9]	**10997.50 (22.37)** [T6]
200x20	11797.40 (52.15)	**11755.82 (57.91)** [T5]	**11753.02 (56.75)** [M3]

Table 3. Simulated Annealing Results — Swap Neighbourhood

Problem	Undirected	Heuristic + Random	Random + Heuristic	Heuristic + Heuristic
20x5	1295.36 (5.19)	1294.62 (5.57) [T6]	**1280.34 (5.23)** [T9]	**1291.34 (7.85)** [M2]
20x10	1609.08 (10.86)	1611.32 (10.22) [M9]	1607.86 (11.04) [M3]	1611.12 (10.72) [T2]
20x20	2340.68 (22.56)	2343.24 (20.03) [T2]	2338.96 (23.32) [T2]	2346.84 (25.16) [T2]
50x5	2737.70 (7.63)	**2734.34 (7.96)** [T2]	**2729.82 (5.63)** [T6]	**2731.90 (6.33)** [M5]
50x10	3116.48 (23.13)	3115.96 (21.62) [M3]	**3103.86 (22.42)** [M6]	3110.80 (20.40) [M3]
50x20	4007.50 (23.25)	4008.46 (24.52) [T5]	4003.22 (19.19) [M5]	4008.78 (18.85) [T2]
100x5	5506.14 (14.65)	**5499.00 (8.04)** [T9]	**5493.76 (0.97)** [T6]	**5494.12 (1.29)** [T4]
100x10	5888.60 (34.58)	5895.82 (31.93) [T2]	**5843.04 (27.54)** [M8]	**5869.80 (27.92)** [T2]
100x20	6562.66 (36.39)	6554.50 (27.33) [T2]	**6526.16 (37.05)** [T4]	**6536.58 (32.03)** [M9]
200x10	11023.30 (38.10)	11012.28 (37.97) [T9]	**10977.42 (31.08)** [T9]	**10984.04 (23.99)** [T9]
200x20	11750.30 (51.57)	**11721.22 (51.10)** [M4]	**11644.56 (47.62)** [T9]	**11674.68 (46.54)** [M8]

Table 4. Simulated Annealing Results — Shift Neighbourhood

nament), and the number gives the Monte Carlo size that was found to give the best performance over a number of test runs. Where the results are highlighted in **bold**, the result was significantly better than the undirected result. Results in *italics* indicate that the result in question was significantly worse. In all cases, significance of any comparisons were confirmed using a t-test.

In general, the shift neighbourhood (operator) gave better solutions than the swap neighbourhood, regardless of whether the neighbourhood was directed or not. Also, the heuristic was able to produce better solutions, though the performance appears to depend upon a number of factors.

The first of these factors was that the heuristic has a much more beneficial effect upon the problems with a large number of jobs — this is not particularly surprising as it could be argued that the smaller problems were not sufficiently difficult for the heuristic to make any impact. Also, the selection pressures that have to be used appear to be quite high — higher than would be used for an EA population for example. This indicates that the use of 'weaker' selection methods, such as the standard implementations of fitness-proportionate and rank-based selection should be avoided (this was confirmed in later experiments).

Finally, how to select the jobs to be moved depended upon the neighbourhood used. The best results for a swap operator were obtained by selecting the one of the jobs at random, whereas for a shift operator, selecting the first job at random when the second job was selected heuristically gave the best results.

Problem	Undirected	Heuristic + Random	Heuristic + Heuristic
20x5	1296.28 (3.08)	**1292.96 (6.67)** [T8]	1295.74 (3.77)
20x10	1620.30 (16.21)	1618.08 (11.96) [T3]	1625.48 (16.74) [T2]
20x20	2358.22 (22.21)	**2348.08 (20.54)** [M5]	2359.16 (24.07) [T2]
50x5	2739.02 (9.35)	**2731.42 (8.19)** [T9]	**2735.80 (8.09)** [T2]
50x10	3128.22 (28.89)	3120.84 (19.60) [M5]	3126.46 (19.96) [T2]
50x20	4031.50 (23.21)	4027.04 (24.48) [T3]	4029.38 (20.40) [T2]
100x5	5517.66 (15.28)	**5503.96 (13.60)** [M5]	**5498.46 (8.54)** [T9]
100x10	5921.44 (31.73)	**5881.18 (30.99)** [T9]	**5902.26 (28.12)** [M8]
100x20	6600.00 (34.02)	**6577.48 (34.64)** [T2]	**6583.22 (33.81)** [T2]
200x10	11027.60 (34.44)	**10990.90 (23.74)** [T8]	**11001.14 (25.87)** [T4]
200x20	11798.70 (44.20)	**11738.06 (48.00)** [T8]	**11745.58 (43.21)** [T2]

Table 5. Threshold Accepting Results — Swap Neighbourhood

Problem	Undirected	Heuristic + Random	Random + Heuristic	Heuristic + Heuristic
20x5	1295.30 (5.09)	1295.40 (4.63) [M6]	**1282.10 (5.18)** [T9]	**1293.02 (6.84)** [T3]
20x10	1604.64 (10.74)	1606.98 (11.66) [T2]	1604.82 (11.10) [T2]	*1609.74 (12.33)* [T2]
20x20	2341.72 (20.15)	2338.74 (20.17) [T2]	2335.14 (21.11) [T2]	2338.76 (20.10) [T2]
50x5	2735.90 (7.23)	2734.44 (9.17) [T5]	**2728.98 (4.56)** [T9]	**2731.68 (6.14)** [M5]
50x10	3115.56 (20.06)	3116.38 (21.57) [M9]	**3104.24 (20.43)** [T8]	3110.48 (19.16) [M3]
50x20	4007.74 (20.17)	4003.50 (20.29) [M2]	**3998.86 (19.19)** [T4]	**4000.38 (19.69)** [T7]
100x5	5507.98 (13.64)	**5499.20 (7.92)** [T9]	**5494.38 (2.75)** [T8]	**5494.68 (1.71)** [T6]
100x10	5889.48 (32.69)	5897.74 (26.39) [M8]	**5842.58 (21.70)** [T5]	**5872.64 (31.10)** [T3]
100x20	6552.14 (32.45)	6549.18 (34.28) [T2]	**6519.38 (27.50)** [T6]	**6530.64 (33.66)** [T2]
200x10	11017.40 (33.09)	11010.24 (30.76) [T8]	**10967.46 (20.10)** [T9]	**10985.76 (26.61)** [T9]
200x20	11736.00 (46.31)	**11703.50 (43.93)** [M3]	**11631.02 (36.63)** [T8]	**11661.32 (41.49)** [M4]

Table 6. Threshold Accepting Results — Shift Neighbourhood

5.2 Simulated Annealing Results

Simulated annealing was implemented, for all problems, with an initial temperature of 12, a final temperature of 0.05 with temperature changes every 100 evaluations according to a logarithmic cooling schedule. This was found to give satisfactory results in the formative experiments in this work. The results for simulated annealing are given in Tables 3 and 4. The results obtained were similar in nature to those obtained for SHC, though simulated annealing did improve solution quality somewhat.

5.3 Threshold Accepting Results

Threshold accepting was implemented, with an initial threshold of 9, and a final threshold of 0, with the threshold changing linearly every 100 evaluations — values that were found to give satisfactory results in formative experiments. The results obtained for threshold accepting were similar in nature to those obtained for SHC, though solution quality was improved (Tables 5 and 6).

5.4 Record-to-Record Travel Results

Record-to-record travel was implemented, for all of the test problems, with an threshold of 5 which was held constant, and which was found to give satisfactory results. The results for RTRT are given in Tables 7 and 8. Again, the results obtained were similar in nature to those obtained for SHC, though generally of higher quality.

Problem	Undirected	Heuristic + Random	Heuristic + Heuristic
20x5	1296.26 (3.63)	**1293.48 (6.57)** [T5]	**1294.72 (5.46)** [M3]
20x10	1619.02 (14.18)	1619.18 (14.58) [M5]	*1632.74 (21.15)* [M4]
20x20	2355.62 (22.71)	2358.20 (24.28) [T2]	*2373.82 (24.46)* [T2]
50x5	2737.66 (11.22)	**2733.04 (7.89)** [M9]	**2732.94 (6.74)** [T3]
50x10	3138.70 (22.01)	**3123.00 (19.41)** [M4]	3132.46 (22.27) [T2]
50x20	4044.50 (26.21)	**4030.00 (26.41)** [M5]	**4032.06 (16.51)** [T2]
100x5	5516.22 (15.10)	**5500.10 (10.93)** [T9]	**5497.38 (7.09)** [T8]
100x10	5917.56 (31.68)	**5883.72 (31.68)** [T8]	**5906.66 (27.92)** [M9]
100x20	6608.70 (32.57)	**6582.32 (33.18)** [M8]	**6589.02 (28.95)** [M8]
200x10	11033.50 (37.67)	**10993.44 (25.49)** [T9]	**10999.36 (29.19)** [M6]
200x20	11819.30 (37.84)	**11761.24 (50.38)** [M9]	**11764.22 (48.41)** [M4]

Table 7. Results for Record-to-Record Travel — Swap Neighbourhood

Problem	Undirected	Heuristic + Random	Random + Heuristic	Heuristic + Heuristic
20x5	1294.14 (6.18)	1294.84 (5.68) [M8]	**1281.32 (7.38)** [T8]	1293.42 (6.14) [M9]
20x10	1609.44 (9.36)	1609.58 (14.52) [T2]	1608.96 (12.39) [M4]	*1613.12 (10.58)* [T2]
20x20	2342.06 (23.12)	2349.18 (24.86) [M3]	2341.70 (22.66) [T2]	2349.62 (25.29) [T2]
50x5	2734.48 (8.22)	**2731.46 (7.26)** [T4]	**2729.50 (5.15)** [T9]	**2730.82 (6.02)** [T4]
50x10	3118.86 (22.65)	3114.46 (23.30) [T2]	**3106.14 (23.40)** [T3]	3113.06 (17.91) [T2]
50x20	4019.06 (22.79)	**4008.44 (18.45)** [T2]	**4005.30 (19.72)** [M5]	**4010.48 (21.58)** [M4]
100x5	5509.24 (15.00)	**5499.40 (8.58)** [T8]	**5493.82 (1.83)** [T9]	**5495.34 (4.16)** [T9]
100x10	5889.70 (33.16)	*5901.92 (31.62)* [T2]	**5838.00 (24.97)** [T9]	**5873.42 (24.88)** [M9]
100x20	6567.06 (37.40)	6568.18 (34.04) [M4]	**6533.82 (27.74)** [M5]	**6540.46 (28.62)** [M3]
200x10	11019.90 (39.49)	**11006.70 (28.39)** [T3]	**10977.66 (25.76)** [T8]	**10984.26 (23.36)** [T8]
200x20	11752.80 (55.88)	11741.34 (48.07) [M3]	**11647.04 (37.13)** [T9]	**11691.72 (41.39)** [M3]

Table 8. Results for Record-to-Record Travel — Shift Neighbourhood

5.5 Evolutionary Algorithm Results

Formative experiments were undertaken to find satisfactory settings for an EA across all of the test problems. This lead to the adoption of a steady-state EA with kill-worst replacement, a crossover probability of 0.3, and a mutation probability of 0.7. Selection of the first parent was performed using rank-based selection, and the second parent was selected at random. The results for the evolutionary algorithm are given in Tables 9 and 10. Unlike the other optimisation techniques, the evolutionary algorithm did not show significant improvements in performance when the neighbourhood was directed, except for the largest problems. Why should this be?

Examination of the results showed that the changes in makespan effected by the directed mutation operator for the EA were of roughly the same size as obtained for the other optimisers considered here. However the standard deviation of the EA results are much higher, which means that the improvements in

Problem	Undirected	Heuristic + Random	Heuristic + Heuristic
20x5	1268.76 (87.63)	1263.42 (90.56) [M5]	1267.66 (89.58) [M3]
20x10	1590.32 (80.20)	1588.32 (81.31) [M7]	1592.44 (80.14) [T2]
20x20	2284.18 (66.64)	2284.58 (73.89) [M6]	2287.56 (71.21) [T2]
50x5	2814.04 (129.00)	2805.04 (132.56) [T9]	2809.58 (129.22) [T2]
50x10	3174.40 (101.20)	3162.00 (98.97) [T8]	3162.66 (104.48) [M9]
50x20	4033.84 (88.94)	4021.42 (82.27) [M5]	4028.04 (86.97) [M4]
100x5	5401.38 (186.55)	5379.84 (190.89) [T9]	5388.70 (190.39) [T3]
100x10	5844.22 (148.34)	5817.38 (145.00) [T5]	5826.96 (145.02) [T3]
100x20	6828.34 (81.23)	6808.98 (92.08) [T3]	6818.50 (98.01) [M5]
200x10	11073.30 (208.54)	11024.76 (217.44) [T8]	11033.40 (216.75) [T3]
200x20	12243.10 (124.37)	12207.40 (131.87) [T3]	**12197.06 (128.21)** [T9]

Table 9. Evolutionary Algorithm Results — Swap Neighbourhood

Problem	Undirected	Heuristic + Random	Random + Heuristic	Heuristic + Heuristic
20x5	1263.06 (90.78)	1262.94 (91.54) [M4]	1260.76 (92.37) [T6]	1262.38 (90.82) [T2]
20x10	1576.00 (78.94)	1580.04 (79.90) [M4]	1575.80 (79.62) [T3]	1584.72 (77.28) [M3]
20x20	2271.08 (69.95)	2272.94 (74.36) [M4]	2270.36 (69.92) [T2]	2275.16 (73.77) [T7]
50x5	2809.12 (130.13)	2810.02 (129.63) [T4]	2798.82 (133.38) [T8]	2803.08 (131.50) [T2]
50x10	3165.16 (103.64)	3161.16 (104.29) [M4]	3145.20 (99.19) [T8]	3155.14 (101.41) [M6]
50x20	4024.14 (82.78)	4015.04 (86.37) [T2]	4009.60 (85.66) [T5]	4009.34 (86.06) [M8]
100x5	5390.96 (185.53)	5394.50 (186.07) [T2]	5372.00 (187.50) [T9]	5378.86 (190.34) [T3]
100x10	5831.74 (144.03)	5832.36 (149.05) [M5]	5800.58 (145.83) [T6]	5813.88 (148.60) [T3]
100x20	6836.50 (86.46)	6817.50 (98.39) [M5]	**6801.18 (90.38)** [T8]	**6804.46 (102.65)** [T2]
200x10	11072.90 (216.02)	11053.84 (219.50) [T8]	**11000.12 (209.69)** [T9]	11030.34 (223.48) [T9]
200x20	12267.60 (133.43)	12252.32 (115.24) [T3]	**12227.52 (113.99)** [T8]	**12222.56 (119.92)** [M9]

Table 10. Evolutionary Algorithm Results — Shift Neighbourhood

performance cannot be said to be statistically significant. As an aside, the much higher standard deviation obtained could be seen to be a reason for not using an EA for this problem: users would prefer a greater degree of certainty of what the final solution will be, and they may chose optimisers that provide this.

6 General Points

Although basic implementations of the optimisers have been used here, it would be interesting to compare optimiser performance. When the optimisers were compared few consistent patterns emerged, though the large variance associated with the evolutionary algorithm results may be a matter for concern. However, all of the implementations of the algorithms here possess considerable scope for improvement, so the choice of any one algorithm over another would be premature at this stage.

Turning our attention to representation, the shift neighbourhood consistently outperforms the swap neighbourhood. Why should this be so? A uninformative answer would be that the shift neighbourhood creates a more correlated landscape, and as all of these techniques work by exploiting such correlations, and directing the neighbourhood has no effect on this, we can account for the observation. But, why should a shift-neighbourhood create a more correlated search space? If example Gantt charts are examined, it soon becomes apparent that there are blocks of jobs that are 'well-meshed' together (ie. have little are no idle time between then). It would appear important that these blocks should be disrupted as little as possible by the neighbourhood operator, or better still, manipulated explicitly. This is precisely what the shift-operator does. In fact, an examination of example Gantt charts showed that the way the jobs in a block mesh together is largely preserved.

In the context of directing the neighbourhood, it was found that selecting both jobs according to the heuristic was not as effective as selecting just one. This is presumably because the former method over-exploits the idle-time information — using idle-time to predict improving moves is only a rule-of-thumb, therefore it is prudent to 'hedge your bets' somewhat, hence the stochastic nature of these algorithms.

The results obtained using the shift neighbourhood indicate that selecting the second job (that which defines the block to be shifted) is more important that selecting the first job (which defines which job is moved out of the way of the shifted block). This is probably connected to the reason why the shift operator is the operator of choice. If the movement of blocks is what is important, then how the block as a whole interacts with the other jobs will be important. As the idle time of the second selected job effectively measures how one end of the block 'fits in' with the rest of the schedule, it can be seen why it results in an effective measure for how blocks should be shifted.

7 Conclusions

On the basis of the results obtained here, the belief that moving jobs that have high idle times would lead to improved solutions appears to be valid. In addition, the method by which this belief is implemented does appear to have an effect, along with the choice of neighbourhood operator.

In addition, and possibly more importantly, the choice of neighbourhood and the directing of the neighbourhood were consistent across optimisation algorithms — a fact that, if noted, is rarely emphasised. The authors would suggest that this is worth noting: experience in mapping domain knowledge to these optimisers is often transferable which allows the user to more easily decide how domain knowledge should be exploited. Thus options can be explored before deciding to use a more complicated optimisation technique such as an evolutionary algorithm which introduces many more choices and parameters that require tuning.

Finally, it is worth mentioning that this idea can be transferred to other problems, such as job-shop scheduling. Also, beliefs can easily be formulated to deal with other optimisation criteria, for example, tardiness. In fact the operations research literature appeared, in the course of writing this paper, to be a good source of such heuristics and users of evolutionary optimisers would do well to turn their attention there.

8 Acknowledgements

We would like to express our gratitude to the Engineering and Physical Sciences Research Council (EPSRC) for their support of Andrew Tuson via a research studentship (95306458). In addition, we would like to thank the reviewers for the references they provided and Chris Gathercole for final proofreading.

References

1. A. Brindle. *Genetic Algorithms for Function Optimization.* PhD thesis, University of Alberta, 1981.
2. G. Dueck. New Optimisation Heuristics: The Great Deluge Algorithm and the Record-to-Record Travel. Technical report, IBM Germany, Heidelburg Scientific Center, 1990.
3. G. Dueck and T. Scheuer. Threshold Accepting: A General Purpose Optimisation Algorithm Superior to Simulated Annealing. *Journal of Computation Physics,* 90:161–175, 1990.
4. C. Hjorring. *The Vehicle Routing Problem and Local Search Metaheuristics.* PhD thesis, University of Queensland, Australia, 1995.
5. J.C. Ho and Y.-H. Chang. A new heuristic for the n-job, M-machine flow-shop problem. *European Journal of Operational Research,* 52:194–202, 1991.
6. J.H. Holland. *Adaptation in Natural and Artificial Systems.* Ann Arbor: The University of Michigan Press, 1975.
7. A. H. G. Rinnooy Kan. *Machine Sequencing Problems: Classification, complexity and computations.* Martinus Nijhoff, The Hague, 1976.
8. S. Kirkpatrick, C.D. Gelatt, Jr., and M.P. Vecchi. Optimization by Simulated Annealing. *Science,* 220:671–680, 1983.
9. S. Minton, A. Phillips, A. Johnston, and P. Laird. Solving Large Scale CSP and Scheduling Problems with a Heuristic Repair Method. In *Proceedings of AAAI-90,* 1990.
10. G. F. Mott. Optimising Flowshop Scheduling Through Adaptive Genetic Algorithms. Chemistry Part II Thesis, Oxford University, 1990.
11. E. Nowicki and C. Smutnicki. A fast tabu search algorithm for the permutation flow-shop problem. *European Journal of Operational Research,* 91:160–175, 1996.
12. I. H. Osman and C. N. Potts. Simulated annealing for permutation flow-shop scheduling. *OMEGA,* 17:551–557, 1989.
13. C. Rajendran and D. Chaudhuri. An efficient heuristic approach to the scheduling of jobs in a flowshop. *European Journal of Operational Research,* 61:318–325, 1991.
14. Peter Ross. Personal communication, 1996.
15. Peter Ross, Dave Corne, and Hsiao-Lan Fang. Improving evolutionary timetabling with delta evaluation and directed mutation. In Y. Davidor, H-P. Schwefel, and R. Manner, editors, *Parallel Problem-solving from Nature - PPSN III,* LNCS, pages 566–565. Springer-Verlag, 1994.
16. E. Taillard. Benchmarks for basic scheduling problems. *European Journal of operations research,* 64:278–285, 1993.
17. T. Yamada and R. Nakano. Scheduling by Genetic Local Search with Multi-Step Crossover. In *The Fourth International Conference on Parallel Problem Solving From Nature (PPSN IV),* 1996.

18. M. Zweben and E. Davis. Learning to Improve Iterative Repair Scheduling. Technical report, NASA Ames Research Centre, Report FIA-92-14, 1992.

19. M. Zweben, E. Davis, B. Daun, and M. Deale. Scheduling and Rescheduling with Iterative Repair. Technical report, NASA Ames Research Centre, Report FIA-92-16, 1992.

Multiobjective Genetic Algorithms for Pump Scheduling in Water Supply

Dragan A. Savic[1], Godfrey A. Walters[1] and Martin Schwab[2]

[1] School of Engineering , University of Exeter, Exeter, EX4 4QF, UK
[2] ERASMUS student, University of Stuttgart, Stuttgart, Germany

Abstract. Cost minimisation is the main issue for water companies when establishing pumping regimes for water distribution. Energy consumption and pump maintenance represent by far the biggest expenditure, accounting for around 90% of the lifetime cost of a water pump. This paper introduces multiobjective Genetic Algorithms (GAs) for pump scheduling in water supply systems. The two objectives considered are minimisation of energy and maintenance costs. Pump switching is introduced as a surrogate measure of maintenance cost. The multiobjective algorithm is compared to the single objective GA, with both techniques improved by using hybridisation with a local-search method.

1. Introduction

Seeking cost reduction and energy savings in water supply by improving the operation of pumps is both an obvious as well as a very efficient possibility to consider. Without making changes to the basic elements of a water supply system, remarkable reductions in operation costs can be achieved by optimising the pump scheduling (PS) problem. The UK Water Research Centre (WRC) [6] estimates the scope for savings for an average system to amount to ten percent of the current pumping cost. Considering that in the UK the overall energy cost for pumping in water supply is about £70 Million p.a., the estimated savings are very high.

The pump scheduling problem can be formulated as a cost optimisation problem which aims to minimise marginal costs of supplying water, whilst keeping within physical and operational constraints (e.g. maintain sufficient water within the system's reservoirs to meet the required time varying consumer demands) [5]. There are two general classes of costs: electrical costs and maintenance costs. Important characteristics of the costs related to the energy consumption are the electricity tariff structure and the civil engineering components of the system, determining the relative efficiencies of the available pumpsets, the head through which they pump and marginal treatment costs. To achieve the aforementioned aim of pump scheduling, an algorithm should be devised to determine which of the pumps available within the system, should be used during which interval of the optimisation period.

Besides energy costs, pump maintenance is another important cost factor in water supply. Although the cost for wear on the pumps caused by the switching of a pump cannot be quantified easily, it can be assumed that it increases as the number of "pump switches" increases. According to Lansey et al. [3], a pump switch is defined as "turning on a pump that was not operating in the previous period". Thus the aim is to limit pump maintenance costs by limiting the number of pump switches.

The issue of optimised pump scheduling has been subject to increased research in recent years. A comprehensive review of optimisation approaches to pump scheduling is given by Ormsbee and Lansey [5]. They considered methods using linear, dynamic and non-linear programming as well as specialised forms of these methods (e.g. mixed-integer linear programming). Their classification is made according to the programming method implemented, the type of system that is to be optimised (e.g. single or multiple tanks), the type of hydraulic model used, the type of demand model used and the resulting control policy. An attempt to develop and use Genetic Algorithms (GAs) for pump scheduling has been reported by Mäckle et al. [4]. They used a binary GA with linear ranking to optimise the pump schedule for a one-reservoir, 4-pump system over a 24-hour period. They concluded that GAs and other evolutionary programming methods can offer a lot of new possibilities for solving pump scheduling problems, in particular by allowing different constraints and cost factors to be considered more easily than with other methods, i.e., linear, dynamic and nonlinear programming.

This paper presents several improvements of the single objective GA and the results of the further investigation into the use of multiobjective GAs for solving the pump scheduling problem. The multiobjective approach used in this work deals with both the energy cost and the pump switching criterion, at the same time. The performance of the algorithm is tested for different demand profiles and additional requirements and compared to that of the single objective GA.

2. Pump Scheduling Problem

The pump scheduling problem presented by Mäckle et al. [4] is used in this work. The system considered consists of one water distribution reservoir which is supplied by four pumps through a single water main. The optimisation period is set to one day as historic patterns of the water demand of an average day are commonly used for pump scheduling. The time interval over which the electricity tariff structure is repeated was also modelled to be 24 hours.

As in many electricity supply systems, the model incorporates a cheap night and a more expensive day tariff. For the problem used in this paper the day-time tar-

Table 1. Pumping capacities of the fixed speed pumps

Pump	Amount of water pumped in one hour [m³]	Amount of electricity used in one hour [kWh]
Pump 1	10	12
Pump 2	30	30
Pump 3	50	44
Pump 4	100	80

iff cost is set to be twice that charged during the night. The period for which the higher day-tariff applies is from 7 a.m. until 8 p.m. The optimisation period was divided into intervals of one hour, i.e. the pumps can be either switched on or off during a particular hour of the day.

There are four fixed speed pumps in the system. As each of the pumps can be run during any time interval, there are $2^4 = 16$ possible combinations of the pumps during each hour of the day. The (constant) pumping capacities of the pumps are listed in Table 1.

With hydraulic problems like the pumping of water, one has to consider friction losses and other head influences to avoid oversimplification. Details about head influences can be found in Hicks et al. [2]. Friction influences are water flow rate, pipe diameter, pipe length, interior condition of the pipe and characteristics of the fluid. The relative friction loss increases with an increase in the amount of water pumped. For this problem it was assumed that the friction loss for flow rates smaller than 100 m³/hour is small and can therefore be neglected. The values listed in Table 2 are for the pump combinations with a flow rate greater than 100 m³/hour.

The maximum storage volume in the reservoir was set to 800 m³ [4]. If the reservoir is filled to this maximum, this volume can be taken from the reservoir and

Table 2. Influences of the pumping head on the electricity consumption

Pumps switched on	Amount of water pumped in one hour [m³]	Amount of electricity used in one hour without head influence [kWh]	Amount of electricity used in one hour with head influence [kWh]
4	100	80	80
1+4	110	92	92
1+3	130	110	111
1+3+4	140	122	124
1+2	150	124	127
1+2+4	160	136	141
1+2+3	180	154	165
1+2+3+4	160	166	182

after taking it out, there is a minimum amount of water left in the reservoir for safety reasons (e.g. fire fighting). The water source is assumed not to have any restrictions. The 24-hour consumer demand profile was adopted from the problem presented by Mäckle et al. [4].

3. The Genetic Algorithms

This section introduces the single objective GA similar to that of Mäckle et al. [4] and the multiobjective GA developed in this work.

3.1 Single Objective Approach

In the pump scheduling case, each pump during a certain time interval is represented by one bit of the string. This means that if the bit's value is zero, the pump it represents is switched off during this time interval. If the value is one, the pump is switched on. A string consisting of 24×4 bits describes the space of possible solutions completely. Thus the overall number of possible solutions to the pump scheduling problem is 2^{96}.

The fitness function for the single objective approach developed by Mäckle et al. [4] aggregates the cost for energy and consumption and penalties for violation of the constraints of the system. The electricity cost C_e is given as:

$$C_e = T_n \sum_{i=1}^{7} C[PC(i)] + T_d \sum_{i=8}^{20} C[PC(i)] + T_n \sum_{i=21}^{24} C[PC(i)]$$

where, i is the time interval of the day, T_n is the night time tariff, T_d is the day time tariff, $PC(i)$ is the pump combination of the time interval i and $C[PC(i)]$ is the electricity consumed in time interval i. Other cost factors can be easily incorporated into the algorithm.

Three system constraints are considered in an approach similar to that of Mäckle et al. [4]. The following constraint violations are incorporated into the current algorithm as penalties:

(1) minimum reservoir level,
(2) the initial water level should be reached in the reservoir at the end of the optimisation period, and
(3) maximum reservoir level (not included directly).

The resulting cost function in terms of energy consumption and constraint violations is:

$$Cost = C_e + \sum Penalty$$

The fitness of individual chromosomes is calculated as the inverse of the overall cost.

3.2 Multiobjective Approach

In contrast to single-criterion models, which proceed to the identification of a so-called "optimal" solution, multicriterion methods provide a choice of trade-off solutions from which a decision maker can select a suitable one to implement. Hence, the aim of the multiobjective (MO) approach is to find a spread of good, trade-off solutions with respect to all objectives. In the case of pump scheduling, the two objectives are the minimisation of energy costs and the minimisation of pump switches. As it is desirable to deal with two objectives separately, i.e. without aggregating them into a single fitness function, the fitness of the individuals has to be determined according to both objectives individually.

The energy cost function used for a single objective case is used for the MO approach as well. The fitness of a chromosome according to the pump switch objective is defined as the inverse value of the number of pump switches a particular pump schedule requires. As a result, a solution requiring fewer switches is considered better according to this objective than a solution with more pump switches.

The multiobjective approach used in this work is based on the concept of pareto optimal ranking as suggested by Goldberg [1]. The method consists of three steps:

(1) finding the non-dominated individuals in the current sub-population,
(2) assigning the same rank to each individual of this pareto-optimal set, and
(3) removing this group of non-dominated individuals from the current sub-population and thus creating a new sub-population.

The aforementioned process is repeated until all individuals of a generation have been assigned to a group. The first set found is ranked one. The rank assigned to the individuals of the following pareto-optimal sets found in this generation is increased by one for every repetition. Finally, all individuals are assigned a fitness score according to their rank.

In addition to providing a set of trade-off solutions, the MO algorithm has to ensure that there are no infeasible members of the final population, i.e. in the set of solutions that is presented to the decision maker (pareto-optimal set). Because of the fact that penalties assigned to an infeasible individual only affect the cost objective, there is a chance of having an infeasible solution in the final pareto-optimal set. To prevent this from occurring a method had to be devised to deny equal probability of reproduction to the infeasible solutions that are in the pareto-optimal set. This was achieved by treating feasibility as an additional objective but with the highest priority. In addition, this objective is used only to distinguish between the top two groups (ranked 1st and 2nd).

4. The Improved Genetic Algorithm

To increase the exploitation features of the GA the option of combining it with a local search method is implemented in this paper. Two different local search strategies based on two different definitions of the neighbourhood of a binary string representing a pump schedule were investigated:

    (1) the neighbourhood is defined as the set of binary strings differing from the given string in exactly one bit, and

    (2) the neighbourhood consists of strings which contain two bits of different values, representing the same pump at different times and are identical in all the other bits.

Regardless of which definition is used, if the new string obtained by searching the neighbourhood represents a better solution, the search proceeds with the new string. The term "better solution" implies different decisions for the multiobjective (MO) than for single objective (SO) approach. In the SO approach the solution newly created by the local search operator is kept if its costs have improved compared to the costs of the chromosome that was altered in the last step of the local search, i.e. the starting solution. Otherwise, the starting solution is kept. In the MO approach the new solution is kept if the two following requirements are met:

    (1) neither the cost nor the number of switches has increased compared to the starting solution, and

    (2) the individual has not become infeasible by the action of the local search operator.

In addition to the local search methods, the progressive penalty assignment was applied in this work. A penalty function that assigns comparatively low penalties at the beginning of the optimisation process and increases penalties during the course of optimisation, was used:

$$Penalty_g = y + \frac{m}{e^{-s \cdot (g+x)} + 1}$$

where, m is the maximum penalty chosen, g is the number of generations produced at the time the function is called, and y, s and x are the parameters of the penalty function. The parameters of the penalty function are experimentally chosen to be $y=0$, $s=0.001$ and $x=-3000$.

5. Results

The two local search methods were applied to both the SO and MO approaches. When applied to the SO approach, the first local search (neighbourhood case "a") found good solutions after fewer generations than the second local search method (neighbourhood "b"). The reason for that may be in the fact that the second method neither changes the amount of water that is pumped nor shifts the pumping work from one pump to another. However, when applied with the MO approach, both local search methods performed equally well. In both cases the GA combined with a local search found good solutions after fewer evaluations of the objective function.

If there is no detailed knowledge of the problem, and the field of application of an optimisation algorithm is not very limited, the algorithm must be able to cope with changes to the model. Furthermore, finding an optimal solution quickly and with a comparatively small computational effort is often required. If the changes to the problem, compared to a previous problem, are not radical, it might be sensible to start the optimisation from solutions found before. In this work, changes in the amount and the profile of the consumer demand and alternatives of the initial water level were considered to be important in this context. Again, both the SO and MO approaches were tested using the aforementioned changes.

5.1 Altering the Demand Profile

The solutions used as an initial population for the test run were produced by running the optimisation algorithm with the original demand profile for the same number of generations and also the same parameter configuration as the runs with the alternative demand profiles. The first alternative demand profile tested differs from the original one only in the distribution of the demand over the day. Hence, the sums of the hourly demands are the same for both profiles and amount to 2570 m^3 in 24 hours. The second alternative demand profile has higher peaks around lunchtime and during the late hours of the day. These have been chosen according to the recommendations given by the Water Research Centre [6]. The overall amount of water consumed is 2670 m^3 and therefore higher than the overall demand of the original profile.

The use of the SO approach with the first alternative demand profile and with the initial population seeded with the final solution from the original run resulted in an improvement over the run which started with a randomly created initial population. Changing the distribution of the demand over the day leads in this case to lower costs for pumping the same amount of water during 24 hours. As expected, the best solution for the second alternative demand profile is slightly more expensive than the one found for the original demand profile, as more water has to be pumped. However, starting from a population seeded with the original solution leads to an improved performance of the GA. The improvement achieved was less pronounced than in the case of the first demand profile because the changes to the system were considerable.

The solutions found using the MO approach and with the initial population seeded with solutions obtained in previous runs, were always well distributed in the cost/switch plane. On average, the number of trade-off solutions in the pareto-optimal set was larger than in the case of solutions obtained with a randomly created initial population. Figure 1 shows ten pareto-optimal sets plotted in the cost/switch plane for runs made with randomly created population (left) and seeded population (right).

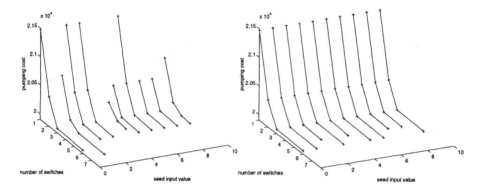

Figure 1. Performance of the MO approach with the alternative demand profile and: randomly created population (left); an initial population seeded with solutions obtained from previous runs (right)

Each of the ten runs was performed using a different random seed number to initialise the search. It is clear from Figure 1 that the coverage of trade-off solutions is better for the runs inoculated by the results from previous runs. The likely explanation for this is that some helpful building blocks were introduced into the gene pool from runs performed for the same system but under slightly different conditions.

5.2 Altering the Initial Water Level in the Reservoir

The initial water level in the reservoir at the beginning of an optimisation period is also the water level that is to be achieved again at the end of the period. Thus, changing the initial water level does not alter the total amount of water that has to be pumped, but the times at which the water must be pumped are changed. Two different cases were considered: (a) lower initial level (75%), and (b) higher initial level (125%).

Getting the initial population for the SO optimisation run by seeding the algorithm with results of similar runs enabled the search to find good solutions for altered initial water level considerably faster than in the case of randomly generated initial population. In the case of the higher initial water level (125%) the algorithm with the seeded initial population even converged on a better final solution (Figure 2). Similar results, in terms of cost, were found for the MO approach as well.

6. Conclusions

In this work the approach to pump scheduling for water supply using the simple, single objective (SO) GA, implemented by Mäckle et al. [4], was investigated and several improvements were introduced. These improvements include:

(1) a hybridisation of the GA search methodology with the local (neighbourhood) search, and

(2) the multiobjective treatment of the pump scheduling problem.

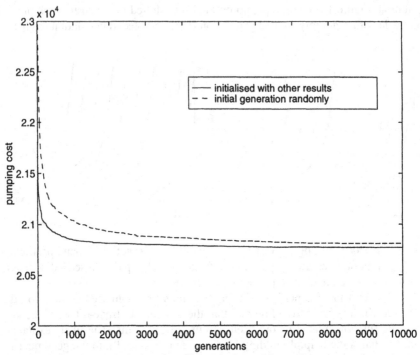

Figure 2. Course of evolution for a SO solution with random or previous-run
initialisation

Two different methods of defining the neighbourhood were investigated and implemented with both the SO and MO approaches. In both implementations, the improvement in GA performance was observed after hybridisation.

In addition to energy costs, the MO approach implemented in this work considered pump switching as an additional objective. The algorithm was tested and improved by the following measures:

    (1) progressive assignment of penalties for constraint violation, and

    (2) introduction of feasibility of solutions as an additional objective (with the
        highest priority).

Based on tests performed, the MO approach achieved very good solutions in terms of
both objectives and generally provides a good set of alternative solutions within the
pareto-optimal group.

Robustness of the two GA approaches was tested by changing the amount
and the profile of the consumer demand and by altering the initial water level in the
reservoir. It was found that even in the case when changes were large (25% change in
the initial reservoir level) the seeding of the initial population with the solutions from
previous runs brought improvements in both efficiency (speed) and quality of the
solutions found.

Acknowledgements

This work was supported in part by the UK Engineering and Physical Sciences Research Council, grant GR/J09796.

References

1. Goldberg, D.E., 1989, *Genetic Algorithms in Search, Optimization and Machine Learning*, Addison-Wesley.
2. Hicks, et al., 1971, *Pump Applications Engineering*, McGraw-Hill, New York.
3. Lansey, K.E. and K. Awumah, 1994, Optimal Pump Operations Considering Pump Switches, *Journal of Water Resources Planning and Management, ASCE*, **120**(1), 17-35.
4. Mäckle, G., D.A. Savic and G.A. Walters, (1995), Application of Genetic Algorithms to Pump Scheduling for Water Supply, *Genetic Algorithms in Engineering Systems: Innovations and Applications, GALESIA '95*, IEE Conference Publication No. 414, Sheffield, UK, pp. 400-405.
5. Ormsbee, L.E. and K.E. Lansey, 1994, Optimal Control of Water Supply Pumping Systems, *Journal of Water Resources Planning and Management, ASCE*, **120**(2), 237-252.
6. Water Research Centre, (1985), *Pump Scheduling in Water Supply*, Swindon, Wiltshire, U.K.

Use of Rules and Preferences for Schedule Builders in Genetic Algorithms for Production Scheduling

K.J. Shaw and P.J. Fleming,

Department of Automatic Control

& Systems Engineering,

The University of Sheffield, Mappin Street,

Sheffield S1 3JD, UK.

Email: shaw@acse.sheffield.ac.uk

WWW: http://www.shef.ac.uk/~gaipp/

Summary

Genetic algorithms (GAs) for problems such as the optimisation of production schedules require large amounts of complex accurate problem information to be included accurately, if optimisation is to be effective. One method of including problem information is the use of an encoding stage, such as a schedule builder, to supplement basic information contained within the chromosomes with data relevant to the manufacturing environment.

The problems of such a representation are explored, when modelling factory decisions , including the use of heuristic rules and preferences. Five schedule builder methods are implemented in the context of a real-life manufacturing example, to compare their effectiveness in improving the genetic algorithm optimisation performance.

1. Introduction

Representation is a vital issue for users of genetic algorithms (GAs) for practical problems. It is essential to the performance of the GA that the accurate information is incorporated, if the GA is to optimise measures based on the modelled application.

This work looks at one particular GA representation in use for optimising production scheduling problems, and issues associated with implementing this representation, for representing rules and preferences used within a real-life manufacturing environment.

The work is presented as follows. The manufacturing problem upon which this work is based is introduced in section 1.1. The issues of problem representation are discussed in section 1.2. Section 2 describes in more detail the genetic algorithm used for this work, and sections 3 and 4 discuss the experiments based on this implementation. The conclusions are presented in section 5, together with suggestions for further work.

1.1 Manufacturing Application Background

The problem in this paper is based on the chilled ready meal problem described previously, Shaw and Fleming, (1996), in which forty-five products have to be assigned to thirteen production lines, given certain constraints, and aiming to optimise three separate costs.

Production of chilled ready meals provides one of the hardest applications for scheduling, (Shaw, 1996). It differs from many standard manufacturing environments in several aspects. The timing of all stages of the production is highly constrained, as orders have a lead-time of hours rather than weeks. Constraints are presented by freshness requirements, stringent hygiene procedures and food safety measures add to the difficulty of producing a feasible schedule, whether it is optimised or not. The manufacturing environment and problem information are constantly changing.

A typical time scale for a production run may be between eight to forty-eight hours from start to finish. The scheduling team will be continually asked to make immediate decisions and alterations to the current schedule based on their expert knowledge of the whole manufacturing process. To meet the higher expectations of consumers, supermarkets extend opening hours to seven days a week, and even 24 hours a day, and continually introduce new product ranges to match trends in food or compete with other supermarkets. Orders must be met satisfactorily and daily to allow fresh food products to be retailed seven days a week.

The problem used in this work is part of a set that have been developed for this research, based on information and data provided by a local manufacturer of chilled ready meals. This is supplemented with information that is general for any food production environment, to provide scheduling problems that incorporate the difficulties presented by such a challenging manufacturing environment.

1.1.1 Problem Description

Provisional orders arrive weekly, and are confirmed each morning for that day's production. The consumer patterns of food purchase are continually subject to changes, making any long-term order forecasting system liable to inaccuracy. There is unpredictability and a need for fast, reactive decision-making within the scheduling process, to ensure that production meets the customers' immediate requirements.

The confirmed orders are assigned to the packing lines to be assembled from prepared ingredients from the cookhouse. The assumption is made that all ingredients are assumed to be available; if this is not the case, manual adjustments are made to the schedule to compensate. Some lines are dedicated to one type of packaging, others are multi-purpose, and each can require varying numbers of staff. Hygiene issues impose restrictions on the relative positions of certain orders on a line; care must be taken that contamination between products does not occur. 'Messy' products may require extra cleaning of the line after their production.

Using these restrictions, the scheduler can build up a provisional schedule assigning jobs to lines. This schedule can require drastic changes because of accidents, sudden changes, or lack of prepared ingredients. Breakdowns and stoppages can occur regularly. Further constraints include staffing considerations, for example, the minimisation of unnecessary overtime, or attempts to ensure that all staff work roughly similar hours. The schedule must also include breaks for staff meals and cleaning. Once orders are completed, there are seven deliveries to go out at regular periods throughout the afternoon. The factory's current main objective is to meet these delivery times. However, it is an important research issue that a developing schedule optimisation system designed for this environment should be capable of meeting more than one objective to allow additional factory goals to be met.

1.1.2 Identification of scheduling problem

Such an environment provides challenging real-life data, offering both suitable complexities and realistic situations upon which a GA-based scheduling system could work. The main application for GA-based scheduling within the factory is the assembly/packaging line assignment. Aspects of this problem, beyond those generally covered in scheduling literature, develop from the use of real-life data from the factory. These include an increasing need for the use of multiple objectives for optimisation and a huge demand for flexible, reactive scheduling.

The placement of the 45 products to the 13 lines is highly constrained, and the assignment of orders to the packaging lines is subject to several restrictions. Strong constraints which must be met include certain lines being configured for particular types of products, adjacency constraints restricting the relative order of products on lines, the requirement that staff have a set number of breaks within a period of production, and time constraints restricting the actual time of day in which certain products can be made, due to the length of certain pre-packaging processes. Weak constraints, which should ideally be met, include the minimisation of several factors. These include the total makespan, the total changeover / maintenance time between jobs, and that all staff have their breaks at an acceptable time.

In addition to such constraints, the factory identified three major objectives that it would like to minimise. These were:

- **Cost 1**- the number of jobs rejected, because they were impossible to schedule given the current constraints

- **Cost 2** - the lateness of any order, with respect to its various delivery times for the seven different depots
- **Cost 3** - the variation between the ends of each of the runs of the 13 production lines.

Cost 1 is not as unlikely as it may first seem. It is certainly the case that the constraints as provided may prevent a particular order from being scheduled at all. In the factory, such constraints would generally be manually relaxed to allow such an order to be placed. However, in extreme cases, given time restrictions for the total production run, jobs might have to be removed from production if accommodating them on the schedule seriously inhibited the production of the other orders.

1.2 Scheduling Problem Representation within GAs

The following section explores the method of incorporating such problem information into a GA. A very simple chromosome plus schedule builder is one common representation when working with genetic algorithms for scheduling. The additional information needed to create the schedules represented by each chromosome is included in a 'schedule builder stage' transforming the chromosome into a schedule which can be evaluated. Subsequent sections discuss the particular use of problem information within the schedule builder.

Opinions conflict as to the most effective way of including knowledge of the domain within the genetic algorithm. It is important to represent the situation of the factory accurately by the inclusion of all necessary information describing it. At some stage, the fitness function, and knowledge of the environment optimised by the GA has to be included. There are varying degrees of direct representation for doing this. Direct chromosomes may have complex configurations, as they are designed to contain as much problem information as possible. Such methods commonly require specially designed operators to ensure that crossover or mutation create legal offspring. However, they can be simply interpreted, as they already represent much of the information directly. Bagchi et al., (1991), recommend this representation.

Indirect chromosomes use simple representations of some of the problem information, for example, a permutation string that simply represents the priority order given to jobs by the scheduler. These strings then have to be interpreted, supplied with additional information, and evaluated in a decoding process, such as a schedule builder. Syswerda, (1991) argues that these are more robust and able to be manipulated more easily.

Bruns, (1993), remarks that 'algorithms that do well across a variety of different classes of problem are usually never the best in any particular problem domain.' This highlights the dilemma of the genetic algorithm scheduler, and indeed for GA users in many other fields, attempting to create a schedule optimisation technique. By creating an excellent solution for a specific problem, there may be a loss of generality for applying the solution to other scheduling environments.

1.2.1 Representation used in this work

There may be no definite consensus on the best form of representation in GAs for scheduling. Indirect representation was chosen for this particularly research, the choice being influenced greatly by the real-life nature of the scheduling problem. A major advantage in the use of the schedule builder was that it allowed extremely flexible inclusion of the environmental data. Given the level of change within the factory, it was vital that this GA representation did not restrict inclusion of changed data in any way. Problem changes can be implemented by a simple alteration to the code contained within the schedule builder, without any need to alter the actual representation or operators used. In addition, the indirect representation seemed less restrictive to the experimental work used in this project on other aspects of GA research.

The main disadvantage of indirect representation is that it may inhibit accuracy of the optimisation process by representing only part of the problem within the population. In this case, the permutations represent only the priorities on the order of the products to be scheduled. However, this disadvantage must be put in context by the fact that the solutions found by the system could only be measured relatively rather than absolutely, given the real-life problem had no known solutions with which to compare ours. At this stage, complete accuracy of optimisation is less of an issue than the ability to find fast, flexible and reasonably accurate solutions to an extremely challenging problem.

Michalewicz, (1996), discusses the inclusion of problem specific knowledge, and in particular, problem-dictated hybridism within the GA implementation, quoting Davis, (1989), stating his belief that real-world knowledge should be incorporated into the GA by operator development or a decoding stage, such as schedule builders, to allow GAs to become useful tools for real-life applications.

2. Genetic Algorithms and Schedule Builders

2.1 Background to Schedule Building

In this section, the use of schedule builders is discussed in more detail. Their purpose is to transform a simple chromosome into a schedule that can be evaluated according to the objectives required in optimisation. Given the permutation, the schedule builder takes it as the priority order in which it must assign the products to the available machinery, according to the rules and constraints provided. The transformation of permutation chromosome into a schedule ready for evaluation provides an essential part of fitness assignment and thus performance of the GA.

Detailed descriptions of schedule builder implementations may be found in several sources developing from the initial suggestion by Davis, (1985); e.g. Whitley, et al., (1989), Syswerda and Palmucci, (1991), Syswerda, (1991), Bagchi et al., 1991. Uckun, et al., (1993), provide a comprehensive discussion, comparing the use of schedule builders against the use of a direct chromosome representation. They comment that a schedule builder extends the search capabilities of the GA into areas not covered by the GA itself, by drawing on the extra information contained within the schedule builder to evaluate the fitness of chromosomes. However, for their implementation, the results are only optimal locally. This is because schedules designed by their schedule builder use optimal process plans whereas 'a globally optimal result might require the use of suboptimal process plans and resource allocations for some job orders.' This issue has subsequently been explored by the Co-evolutionary Distributed GA, (McIlhagga, Husbands and Ives, 1996). It is useful to note this that combining what appear to be two locally optimal components of problem may not actually provide a globally optimal solution to the problem. This will be discussed in context to this problem in section 3.1.

More recently, the choice of rules used within the schedule builder is examined. Dorndorf and Pesch (1995), describe the use of the Giffler and Thompson (1969) algorithm for assigning operations to machines. This solves the conflict of two (or more) operations competing for the same machine randomly - the authors suggest the use of one of twelve priority rules to resolve these conflicts. Further work on schedule builders includes a method using genetic programming, to evolve the best scheduling heuristics to use within the schedule builder (Langdon, 1996). Also relevant to this area is the 'Evolving Heuristic Choice' (EHC) method (Fang, Ross and Corne, 1994), which includes heuristic rules to be used in the schedule builder as part of the chromosome encoding, allowing them to evolve alongside the schedule, and produces far superior results on a set of benchmark problems. Attention is drawn to the effectiveness of including a degree of hybridisation within the GA process, by such a technique.

2.2 Genetic Algorithm Implementation

The genetic algorithm that was initially implemented to solve this problem was designed to be as simple as possible, for reasons discussed below.

The GA used a generational population of individuals that were permutations of the integer values (1,2,..,45), each element within the permutation representing one product. The population size was set at fifty individuals, over forty generations. Fitness evaluation took place after the individuals had been coded into complete schedules by the schedule builder, which is discussed in more detail below. The objective values were calculated as a weighted sum of the three costs provided, for later development as a multi-objective optimisation GA (Shaw and Fleming, 1996). The Order Crossover, and Swap Mutation Operators were used, these being operators that had been designed to create legal offspring from permutation strings, and which had been shown experimentally to work well for this particular problem.

3. Methods of Line Assignment

As the GA individuals are permutations, it is the role of the schedule builder to take each element of the permutation in turn to assign the order represented by that element to a suitable and available line. The method of selecting between more than one available line for a particular job must be addressed. As may be the case when working with real-life problem data, the explicit information required for setting precise rules was not readily available from the factory. Line assignment in the factory takes place by the scheduling team's experience, by rule-of-thumb, or according to an unquantifiable combination of factors at the time of scheduling. Further consideration was needed upon the method of line assignment given a choice of available lines, both from the theoretical point of view - which line would provide with the best schedule for a global optimal solution, and the practical - whether the choice was sensible and within the scope of the factory's own experience.

The simplest method of such decision making, given the lack of any alternative information was simply to pick a line at random from those available. However, potential problems with this were indicated for a case in which several lines might be available simultaneously, giving so many potential mappings of chromosome to schedule that a certain element of unreliability in the assignment of a fitness to the permutation would result. Any given individual might have a whole range of fitnesses available depending on the random selections of lines made. The effects of this variation in fitness assignment from generation to generation would need closer examination to see the effective on the overall performance of the GA. Clearly, a method that allowed a more informed selection from the available lines was required, and alternative methods of informing this choice are explored below.

3.1 Heuristic Preferences

One simple method of choosing a line from a given set is the use of heuristic rules to make the final decision. Clearly, any such rule would provide a one-to-one mapping of fitness to individual. Common examples of such rules are : 'select the line that finishes soonest ' or 'select the line requiring the minimum changeover'. Whilst this is useful in modelling accurately the events taking place in the factory, it is perhaps necessary to question its effectiveness from the optimisation point of view. Does using such a heuristic actually limit the search of possible schedules too much? Although a heuristic is used in the factory, it might not necessarily the most efficient method to help find an optimal schedule. It may be convenient to use, as the data is already available, or it may be a common rule-of-thumb in other factories. A heuristic may limit the use of the GA by fixing the elements too rigidly and not allowing sufficient exploration of the available schedule space. One example given was that by using a 'choose the line that finishes earliest' heuristic, the rule is implicitly encouraging the optimisation of early finishing times, which might not necessarily always be desirable, especially with the emphasis on trade-offs between conflicting objectives (Shaw and Fleming, 1996), requiring that all objectives be treated equally, without an implicit bias towards one particular objective. If this method were to be used, it would be necessary to choose between the heuristic that best represents the action taken in the factory in real life, or choosing between a large selection of available heuristics to see which allows us most effective use of the GA as an optimisation tool. The actual search for a best rule to use in this situation has been explored previously. (Fang, Ross and Corne, 1994; Dorndorf and Pesch, 1995; Langdon, 1996) This work aims to examine the effectiveness of using rules against an alternative method.

For this particular example, examination of the problem data and factory practices provided one practical heuristic which could be included in the problem at this stage, that of choosing a line with the minimum changeover period. Although there are many other heuristics well known in the literature on scheduling, data collection and factory preference elicitation beyond the scope of this work would have been necessary to include other such rules at this stage.

3.2 Pre-expressed Preferences

A less rigid decision could be made by the use of pre-expressed preferences of which line to use for any given product. By creating a matrix that explicitly expressed a value of preference for each assignment of a product to a line, the choice of lines could be made by simply from the line available with the highest expressed preference. These preferences could be fixed throughout the run of the problem, or again, could be adaptively altered according to the situation on the shop floor (for example, if a machine is starting to fail during a shift, preferences might be expressed to allow that line to 'go easy' when possible). As a trial, these preferences are expressed on a scale from 1 to 10, or perhaps in simpler terms of 'good', 'neutral' and 'bad' (mapped 1-3).

Further investigation into the use of these preferences may take place to see how exactly they need to be expressed. If each product/line preference value is expressed as a unique value, there will only ever be one possible outcome when a choice between lines has to be made. This allows us to provide a one-to-one mapping for the chromosome-schedule interpretation. However, this could not only be fixing the problem too rigidly, limiting the problem to the extent that the potential for search possible schedules is over-constrained. Use of integer preferences, where some lines may have the same preference for a given product, would reintroduce a certain random element in their one-to-several mapping of fitness to chromosome, where there may have to be a further choice between two lines with the same preference. However, the extra level of constraints would provide a narrower choice than in the random schedule builder. Experimentation should show how far this choice can be relaxed.

4. Experiments on the performance of schedule builder

Given the possible rules that might be used for the allocation of jobs between multiple available lines, experiments were conducted to explore their performance within the GA optimisation process. The performance of the following schedule building techniques were compared:

	Technique	Notation for Preferences Used	Implementation
SB1	'Random Line' Schedule Builder	pref_0	Given a choice of lines when assigning a job, choose a line at random
SB2	Preferences - unique	pref_random	Assign unique preferences for each job to line assignment
SB3	Preferences - 1 - 10	pref_10	Assign a preference on the scale of 1 - 10 for each job-to-line assignment
SB4	Preferences 1 - 3	pref_3	Assign a preference on the scale of 1 -3 for each job-to-line assignment
SB5	Problem Specific Heuristic	maintenance	Given a choice of lines when assigning a job, choose the line with the minimum maintenance time

Table 1 - Schedule Building Techniques under comparison

SB1 is described above in section 3, SB2 - SB4 are variations of the method discussed in section 3.2. The problem specific heuristic, discussed in section 3.1, was chosen as being the best example of such a rule for this problem.

Tests on the capabilities of the schedule builders were planned as follows:

4.1 Experiment 1 - Level of variation in fitness assignment

It has already been suggested that it is undesirable to have a high level of variation in the chromosome-to-fitness mapping. Each schedule builder was run 500 times on one specific chromosome, to provide sampled distributions of the fitnesses created by each method for an individual chromosome. This allows a profile of the variation of fitnesses provided by each method to be shown, by indicating the distribution of possible fitnesses that could be assigned to each individual by the schedule builder.

4.2 Experiment 2 - Performance within the GA

The actual performance of each schedule builder within the GA was put to the test by comparing the performance of ten runs of a standard, weighted sum scheduling GA (Shaw and Fleming, 1996) for each schedule builder. The GA used 50 individuals, for 40 generations, using order crossover (Davis, 1985), splice mutation (Fox and McMahon, 1990), which were found in previous tests to work well on this problem.

4.3 Results

The results of the above experiments are presented below.

4.3.1 Experiment 1

Experiment 1 examined the actual variation in fitness assignment to each individual for the different schedule builder methods. Figure 2 shows the ordered fitnesses produced by 500 repeated runs of each schedule builder on one single individual. Clearly, the random method (pref_0) provides a wide range of fitnesses for the same individual, with a difference of over 0.3 between its best and worst. pref_3 gives a similar amount of variation, with somewhat less extreme values at the top of the range of fitnesses. Pref_random and Maintenance provide the same value throughout - a one-to-one mapping of fitness to chromosome. Pref_10 also shows variation across the range of fitnesses.

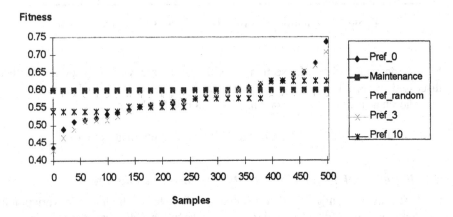

Figure 1- Graph of the variation in distribution of fitnesses assigned to one individual by each method.

4.3.2 Experiment 2

Experiment 2 applied the schedule builders' capabilities to the problem by running each within a genetic algorithm.

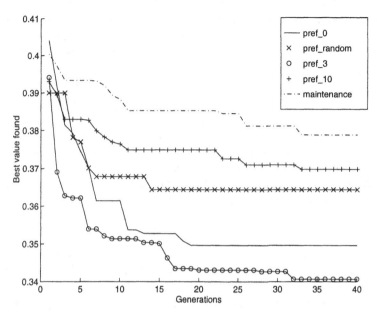

Figure 2 - Plot of mean performance over five runs of each schedule builder within a genetic algorithm

The mean performances of each of the five runs are shown averaged for each schedule builder in Figure 2. From this, it can be clearly seen that pref_3 performs best, followed by pref_0 (the random 'no additional information' technique), and so the other three methods actually impair the performance of the GA compared to its original 'random choice' method.

It is interesting that the priority rule method ('maintenance') is actually the worst of all the methods shown. This rule-of-thumb is perhaps the most realistic representation of a method that might be used to make a decision in the factory, but as can be seen from this, it is certainly not the most effective way of doing so. An approach that may seem common sense in the factory - to select the line with the shortest maintenance time to get the production of the next order running as soon as possible - damages the overall optimisation of the schedule.

The pref_random and pref_10 techniques do not fare very much better, suggesting that again, the rigid prescription of preferences (whether unique, in the case of pref_random, or with only a slight amount of flexibility, as in pref_10) may be too rigid in constraining the problem to finding sub-standard solutions - without any flexibility, the schedule builder may not be capable of getting near an optimal solution.

The random GA, pref_0, actually performs better than these other methods, indicating the degree to which care should be taken in choice of preference rules within the schedule builder - that the GA can perform so well given the amount of variatio7n contained within it is encouraging for supporters of GAs in itself, but that it outperforms methods intended to improve upon the random choice of lines, by limiting such variation in assignment, gives us a particularly cautionary example.

Finally, the best performance is given by pref_3. This method limits the variation within the schedule builder somewhat, but by preserving an element of this variation, allows the GA to find far better solutions. We would suggest that, as we do not know the preference matrices that would best enhance the GA's ability to find an optimal solution (and indeed, if we are creating these preference matrices from real-life information, we are not necessarily permitted to do so!), allowing a little variation within the schedule builder is better than rigidly prescribing preferences which, as can be seen, do little to help the optimisation performance.

5. Conclusion

The use of an interpretation stage within a GA, such as a schedule builder, for a large problem based on real-life information depends on both the accuracy and degree of consistence in assigning fitnesses to individuals. This work explores issues involved in including such information for a real-life manufacturing problem.

Whether the schedule builder representation should be used at all is still something of an undecided issue. In this work, such representation has offered many advantages, such as flexibility for fast-changing problem data, portability, and ease of use for experimental work. Some of the problems of rule selection within the GA might have been avoided by using a more direct representation of the problem, and this would offer scope for future work. However, it has been an aim of this work to develop schedule builders that allow flexible incorporation of such data, and to explore new types of preference representation, without losing any accuracy of the schedule optimisation.

Within the framework of schedule building, heuristics may be the solution most accurately reflecting behaviour in factory decision making, but the results bring into question the actual effectiveness of certain heuristics when working with optimisation. Accurate reflection of factory decision making may impair genuine optimisation rather than aid it if there is no influence in the choice of the heuristic towards the actual goal of the optimisation. Choosing a rule because, 'it's always done this way,' may maintain realistic representation of the factory, but will not necessarily be the ideal for optimisation of costs. Methods such as Langdon, (1996), Fang et al. (1994), may allow the selection of a heuristic that is beneficial to the optimisation process.

Clearly, there is much scope for modelling and inclusion of alternative heuristics from the factory, which may provide better performances within the GA, and this might be an area for future work. However, it is worth noting that it is the case when working with real-life problems that the obvious choice of heuristic may be irrelevant, or the necessary data or information for its implementation may not be readily available.

Preferences allow us to explore the degree of constraints that might be included in the schedule builder before the search either becomes too narrowly constrained or too randomly implemented. It allows information from the factory to be included and incorporates a less rigid control than a heuristic might. However, the use of this method would depend on the ability to capture suitable knowledge and have it expressed as a numerical preference at this stage. When using the method in practice, it may be better to choose a form in which preferences more broadly defined (pref_3) than one which assigns rigid values for preferences (pref_random) - given that this makes sense in the context of the problem, and does not conflict with any fixed requirements in the factory.

We would suggest that the idea of use of a certain amount of flexibility within the schedule builder, as demonstrated by the performance of pref_3, seems to be an interesting area for further exploration of improving the performance of any genetic algorithm which includes a degree of uncertainty in its evaluation of the population individuals' fitnesses, as is the case here.

For both of the other methods (heuristic preferences and pre-expressed preferences), we should note that although both might constrain the search far more than we would like for pure global optimisation, it may be that, given the practical time constraints and factory requirements, it is better to use these modifications to give a reliable search that is less optimal in its treatment of the problem space than one that is unconstrained in its exploration of the problem space but is unstable and unpredictable (the random method, SB1).

More work on these modifications may suggest particular applications in which they may be useful, and may indicate how further work can allow such problems to be effectively searched by GAs in future.

The authors gratefully acknowledge EPSRC grant (GR/K31343). The authors also gratefully acknowledge the co-operation of Pennine Foods with this work. Pennine Foods is a Northern Foods Company.

6. References

Bagchi, S., Uckan, S., Miyabe, Y., Kawamura, K., 1991. Exploring Problem-Specific Recombination Operators for Job-Shop Scheduling, Proceedings of the Fourth International Conference on Genetic Algorithms. Morgan Kaufmann Publishers

Davis, L., 1985. Job Shop Scheduling with Genetic Algorithms, Proceedings of an International Conference on Genetic Algorithms, 1985.

Davis, L., 1989. *Adaptive Operator Probabilities in Genetic Algorithms,* Proceedings of the Third International Conference on Genetic Algorithms and their Applications, 1989, pp. 60 - 69.

Dorndorf and Pesch, 1995. *Evolution Based Learning in a Job Shop Scheduling Environment*, Computers and Operations Research, Vol. 22, Issue 1, pp. 25 - 40.

Fang, H. Ross, P. and Corne, D., 1994. *A Promising Hybrid GA/Heuristic Approach for Open-Shop Scheduling Problem*, ECAI 11th European Conference on AI, Wiley.

Fox and McMahon, 1990. *Genetic Operators of Sequencing Problems*, Foundations of Genetic Algorithms, ed. G. J. E. Rawlings, 284 - 301.

Giffler B., and Thompson G. L., 1969. *Algorithms for solving production scheduling problems*, Operations Research 8: 487 -503.

Langdon, W., 1996. *Scheduling Maintenance of Electrical Power Transmission Networks Using Genetic Programming*,, GP-96 Conference, John Koza (ed.), Stanford Bookstore.

McIlhagga, Husbands and Ives, 1996. *A Comparison of Optimisation Techniques for Integrated Manufacturing Planning and Scheduling*, Proceedings of Parallel Problem Solving from Nature IV.

Michalewicz, Z., 1996. *Evolutionary Computation; Practical Issues*, International Conference on Evolutionary Computation, ICEC '96. pp. 30 - 39.

Shaw, K.J., and Fleming, P. J., 1996. *An Initial Study of Practical Multi-Objective Production Scheduling Using Genetic Algorithms*, UKACC International Conference on Control '96, pp 479 - 485, IEE.

Shaw, R., 1996. *Extending the Shelf Life of Chilled Ready Meals*, Developments in Meat Packaging, ed. Taylor A. A., ECCEAMST, Utrecht, The Netherlands.

Syswerda and Palmucci, 1991. *The Application of Genetic Algorithms to Resource Scheduling*, Proceedings of the Fourth International Conference on Genetic Algorithms, pp. 502 - 508.

Syswerda, G., 1991. *Schedule Optimisation using Genetic Algorithms*, Handbook of Genetic Algorithms, ed. Davis, L., Van Nostrand Reinhold.

Uckun, Bagchi, Kawamura and Miyabe, 1993. Managing Genetic Search in Job Shop Scheduling, IEEE Expert, October 1993, pp. 15 - 24.

Whitley, D., Starkweather, T., and Fuquay, D., 1989. *Scheduling Problems and Traveling Salesman; the Genetic Edge Recombination Operator*. Proceedings of the Third International Conference on Genetic Algorithms, pp 133 - 140.

Novel Techniques and Applications
of Evolutionary Algorithms

A Voxel Based Approach
to Evolutionary Shape Optimisation

Peter Baron[1], Robert Fisher[1], Andrew Sherlock[2], Frank Mill[2], Andrew Tuson[1]

[1]Department of Artificial Intelligence, University of Edinburgh, 80 South Bridge, Edinburgh EH1 1HN. Email: {peterba,rbf,andrewt}@dai.ed.ac.uk

[2]Manufacturing Planning Group, Department of Mechanical Engineering, University of Edinburgh, King's Buildings, Mayfield Road, Edinburgh, EH9 3JL. Email: A.Sherlock@ed.ac.uk, F.Mill@ed.ac.uk

Abstract. Shape optimisation is a hard problem from the field of Mechanical Engineering with the potential for significant cost savings if successfully performed. In the past, evolutionary optimisation approaches have proved successful. In those studies, a form of the shape was assumed, and its parameters optimised. An alternative is a voxel (N-dimensional pixel) based representation, which makes no such assumptions about the form of the solution, and allows the user to add domain knowledge if desired. This paper outlines a preliminary investigation into this approach and shows that the objections to this approach in the literature can be overcome if care is taken over the design of the operators.

1   Introduction

Shape optimisation within constraints is a hard problem from the field of Mechanical Engineering. The objective is to design a shape which best satisfies some predetermined goal whilst at the same time maintaining some property of the shape within a constraint, or perhaps even a set of constraints.

Previous work has applied genetic algorithms [5] to shape optimisation problems with encouraging results. Using engineering software packages to evaluate the suitability of a given design, successful shapes have been evolved — examples include [8,6]. However, previous work on evolutionary shape optimisation has primarily been concentrated around parametric representations of structural design and shape optimisation problems.

The investigation detailed here evaluates the use of a voxel (N-dimensional pixel) based representation instead, within which the shapes being optimised are represented as a series of binary 0's and 1's. This approach has the advantage that it can describe any topology, but it also has its drawbacks. The question is: are the drawbacks sufficient to prevent this approach being useful?

This paper will outline the case for such an approach, and describes an implementation of a basic genetic algorithm, and an extended version. A summary of the results obtained for a simple shape optimisation problem is then given, from which we conclude that the potential problems of the voxel representation that have been suggested in the literature can be circumvented if suitable operators are designed and used.

2 The Case for a Voxel-Based Approach

Traditionally, the representational approach used in structural design has been parametric — a form of the solution, for example as a set of splines, is derived or assumed and the parameters associated with this form are optimised. However, parametric approaches do make strong assumptions about the form of the solution, which may not necessarily correspond to the optimal solution. In addition, it is not always straightforward to devise a good parametric form for some problems.

Taking beam design as an example, holes in the shape of the beam are entirely possible and even probable if some of the mass of the beam is occupying a low-stress area — a parametric representation can only create holes where the user is expecting them to be required and has defined the appropriate parameters.

A report by [7] has discussed the possible use of voxels as a representational approach. As this encoding can describe any shape, it can deal with the situation above, and makes no assumptions about the form of the final solution. Furthermore, a voxel based representation, by virtue of its directness of representation allows domain knowledge to be easily added, and to the level felt appropriate by the designer. Two examples will help illustrate this.

First, areas of the voxel representation can be fixed to be permanently on or off; this allows the designer to prohibit material from being placed in locations where it is not desired. A second example lies with the ease in which existing designs can be utilised by the system — all that is required is for the initial design to be digitised and the bitmap used to initialise the population of the genetic algorithm.

However, [8] argues that using a binary voxel representation leads to the following problems:

- The long length of the chromosomes (> 1000 bits).
- The formation of small holes in the shape.
- No guarantee that the final shape produced will be smooth.
- Even if the parents represent a valid shape, the children will not necessarily be valid.

This paper will evaluate whether these objections constitute a problem in practice.

3 The Optimisation of the Cross Section of a Beam

One of the problems encountered in shape optimisation, no matter what approach is taken, is interfacing the optimiser with a suitable evaluation package. For example, in wing optimisation, the wing shape has to be smooth, else the

CFD (Computational Fluid Dynamics) package will act strangely: either return-ing negative drag, or resulting in the program crashing. Problems involving Finite Element Analysis (FEA) evaluation packages are somewhat better behaved, but interfacing is still not trivial, and the calculations do take some time.

The shape optimisation of a beam cross-section is the problem considered in this study. Evaluation of the candidate cross-sections was made using bending theory for symmetrical beams, considering only normal stresses [4]. This is a greatly oversimplified model, but sufficient to test whether the potential prob-lems with a voxel representation outlined above do pose a problem in practice.

Each candidate solution can thus be represented as a 2-D grid of voxels, with the optimisation objective being to minimise the mass m of the beam whilst ensuring that at all points (ie. at all voxels) in the beam the normal stress does not exceed a maximum stress (σ_{max}), which is a constant that is determined by the material to be used. This maximum stress constraint is given by Equation (1):

$$\sigma_{max} \geq \frac{-My}{I} \ for \ all \ voxels \tag{1}$$

where M is the bending moment, y the distance of the voxel from the neutral axis, and I is the second moment of area of the cross-section. The neutral axis of a shape is defined as a line which passes through the centroid of mass of the shape. As the voxels are of uniform size and density, the centroid can be found by taking the average of the positions of all the occupied voxels; also, for a symmetric beam the neutral axis would be horizontal.

The bending moment M is a constant determined by the loading on the beam. The mass m of the beam is proportional to the area of the cross-section and hence the number of voxels turned on. The second moment of area is given by Equation (2):

$$I = \sum_{i=0}^{all \ ``on" \ voxels} y_i^2 \ dA \tag{2}$$

where y_i is the distance of the ith voxel from the neutral axis and dA is the area of a voxel (as we are dealing with a cross-section).

The optimum cross-section for a beam using this evaluation can be deduced to being a flange at top and bottom of the design domain. For the experiments described here, the following values of the constants were used: $M = 2 \times 10^6$ Nm, $\sigma_{max} = 100$ MPa, and a beam of dimensions 320×640mm was considered.

In practice, this would correspond to an I-beam, but that also requires a web to connect the two plates of the beam together. In a full calculation with shear stresses, the web would arise so to counteract this additional stress. However as shear stress is not represented in this problem, a connectivity requirement in the form of a repair step was added, whereby all pixels must be connected (by a 4 connect rule) to a seed pixel in the centre top edge of the beam. In addition, a straight web was enforced before the connectivity repair step. This was found, in formative experiments, to prevent the formation of a crooked web (as the physics model used does not prevent this), and improve slightly the results obtained.

4 A Basic Genetic Algorithm Implementation

This study builds upon the following basic implementation of a genetic algorithm. The encoding digitises the shape as a 32 × 64 grid and represents this as a 1-dimensional string of 2048 bits. Standard two-point crossover and bit-flip mutation were used to operate on this encoding. After the application of each operator, pixels that were not connected to the seed pixel were set to zero. An unstructured, generational population model of size 20 was used, with rank-based selection with selection pressure 1.7.

Strings are initialised randomly to a set 'density' percentage, the higher this value, the more voxels are initially turned 'on'. For this study, this was set to 70%. All initial population strings must pass the validity checks used by the evaluation function: the number of active voxels must be non-zero; the second moment of area must be greater than 1×10^{-12} (goes to 0 if insufficient active voxels); and the fitness must be greater than 0.0001.

4.1 The Fitness Function

The fitness function was designed to minimise the area of the beam (number of active voxels) within the maximum allowed stress, with a penalty value being applied to any solution which broke that constraint. A small additional factor, $\frac{1}{S}$, was included in the fitness calculation based upon the maximum stress point in the beam. This had the effect of causing any valid solutions to continue to evolve towards better solutions (in this case a beam which minimises area and minimises the maximum amount of stress present). The fitness, F, is given by Equation (3) below:

$$F = \frac{1}{V - \frac{1}{S} + k \times max\{(S - \sigma_{max}), 0\}} \tag{3}$$

where V is the number of active voxels (ie. beam mass), S the highest stress (calculated using Equations (1) and (2)) felt by any of the pixels, σ_{max} the maximum allowed stress, and k a constant which can be adjusted to vary the weight of the penalty associated with the maximum stress constraint (in this study it was set to $k = 5.0 \times 10^{-5}$).

4.2 Results Obtained

The basic genetic algorithm was found to give disappointing results. When allowed to run to convergence (over 2000 generations), the shapes produced, though recognisable as approaching an I-beam shape, were highly irregular and possessed small holes. Figure 1 illustrates this by showing the 4 best solutions from ten genetic algorithm runs. Therefore it appears, at least for a simple genetic algorithm implementation, that the potential problems described in Section 2 do manifest themselves.

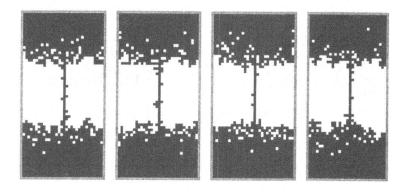

Fig. 1. Typical End-of-Run Results Obtained by the Basic GA

5 An Improved Genetic Algorithm Implementation

The performance of the basic genetic algorithm was disappointing, however, many successful applications of the genetic algorithm use domain specific operators [3]. Knowing the nature of the problems that arise with this approach, would it be possible to design operators to overcome them? With this end in mind, the basic operators were replaced in order to improve the following:

- Crossover effectiveness.
- Removal of holes and isolated pixels.
- Removal of rough edges.

All of the operators also address one weakness in the basic genetic algorithm implementation: the encoding of a 2-dimensional problem as a one-dimensional string. The modifications made will now be described in turn, and results of experiments to evaluate their effectiveness summarised.

5.1 A Mutation Operator for Smoothing

A mutation operator for smoothing was then devised. An area of x/y size ranging from 2 pixels to 1/4 of the dimensions of grid was randomly selected. The most common value for the pixels in the area selected was then found, and then written to all of the pixels in that area (Figure 2).

Examination of the results obtained confirmed that this can remove isolated pixels with great success. With this as the only mutation operator, the shape was optimised to two near perfect horizontal bars at the vertical extremities (an I-beam) after 1000 generations.

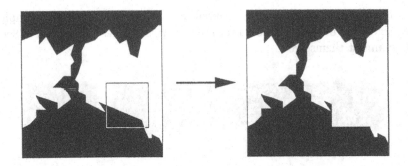

Fig. 2. The Smoothing Mutation Operator

5.2 Two-Dimensional Crossover

Another problem encountered in representing what is a 2-dimensional optimisation problem, as a one-dimensional chromosome, arises with *linkage*. In a 1-D encoding, voxels that correspond to spatially close points on the actual shape can be far apart on the string. Therefore, use of crossover can more easily disrupt building blocks such as one part of the shape being of high fitness, because the bits that correspond to it are spread across the string.

Such a situation has been encountered before in the use of a genetic algorithm to solve the source apportionment problem [1,2]. That investigation found that representing the problem as a 2-D matrix, and devising a crossover operator (UNBLOX) that swapped a 2-dimensional section between solutions lead to improved performance (Figure 3).

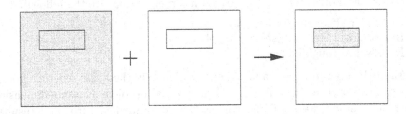

Fig. 3. The UNBLOX Crossover Operator

When UNBLOX was implemented for this problem it was able to produce a recognisable I-beam after 500 generations — a large improvement in performance over the one-dimensional two-point crossover operator.

5.3 2 × 2 Area Mutation Operator

This operator acts on a 2 × 2 area of the chromosome array, and only modifies the contents if at least one voxel in the chosen area is turned on and one other voxel is turned off. This so so that this mutation operator would work well on surfaces. If so, standard bit-flip mutation is applied to each voxel in the area. This operator was devised for three reasons: first, the belief that as we are tackling a 2D problem, the operators should reflect this; second, a 2 × 2 area was thought sufficient to eliminate many of the loose pixel/wobbly line problems found with the basic genetic algorithm; third, applying the operator only where at least one pixel is present gives a good probability of a worthwhile modification being found. This operator can also be thought of as having Lamarkian characteristics.

Use of this operator greatly increased the rate at which excess material is chopped away, whilst the ability of the genetic algorithm to continue to find improvements after convergence was also improved. In addition, the final form of the I-beam was found to be much cleaner than that obtained by the basic genetic algorithm.

5.4 The Final System

After the formative evaluations of each of the new operators above, it was necessary to see if these operators would work effectively in combination. Therefore, the genetic algorithm was run with the following settings: $p(UNBLOX) = 0.3$, $p(2 \times 2 \, mutation) = 0.125$, $p(Smoothing \, mutation) = 0.125$ (all probabilities are per-string); where each operator is applied sequentially to the string with these probabilities. The probability of the 2 × 2 mutation operator was increased by 0.0005 per generation to a maximum of 0.4 as this was found to improve the quality of the results obtained slightly. All of the other genetic algorithm settings were left unchanged from the basic genetic algorithm.

Fig. 4. Typical End-of-Run Results Obtained by the Improved GA

Figure 4 shows some typical end-of-run results, which are near-perfect I-beams, with no holes — a noticeable improvement over the basic genetic algorithm (Figure 1).

Plots of the fitnesses against generation for the basic and improved genetic algorithms are given in Figure 5 (these figures are an average over 10 runs). As can readily be seen, the new operators have dramatically improved the performance of the genetic algorithm, finding a better quality solution more quickly than the basic genetic algorithm, thus vindicating our approach.

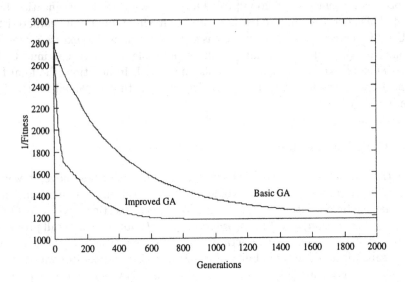

Fig. 5. Fitness Plots for the Basic and Improved GAs

6 Conclusion

This paper has highlighted the possible utility of a voxel-based representation for shape optimisation and has shown, contrary to previous arguments in the literature, that this approach is suitable for evolutionary shape optimisation when used in conjunction with suitably designed operators. The length of the strings, and the validity arguments were found not to be problems in practice — the augmented genetic algorithm was found to be able to discard quickly poor or invalid solutions in favour of good ones. It would appear that, on the basis of these results, a voxel-based approach to shape optimisation would be viable for real problems, especially when you consider that the population has been randomly initialised — which would not often be the case when tackling a real problem. Work applying this approach to real problems, with a view to a comparision with parametric methods, is currently underway.

7 Acknowledgements

Thanks to the Engineering and Physical Sciences Research Council (EPSRC) for their support of Andrew Sherlock and Andrew Tuson via studentships with references 95303677 and 95306458.

References

1. H. M. Cartwright and S. P. Harris. Analysis of the distribution of airborne pollution using genetic algorithms. *Atmospheric Environment*, 27:1783–1791, 1993.
2. H. M. Cartwright and S. P. Harris. The Application of the Genetic Algorithm to Two-Dimensional Strings: The Source Apportionment Problem. In Stephanie Forrest, editor, *Proceedings of the Fifth International Conference on Genetic Algorithms*. San Mateo: Morgan Kaufmann, 1993.
3. L. Davis, editor. *Handbook of Genetic Algorithms*. New York: Van Nostrand Reinhold, 1991.
4. J. M. Gere and S. P. Timoshenko. *Mechanics of Materials 2e*. Brooks/Cole Engineering, 1984.
5. John H. Holland. *Adaptation in Natural and Artificial Systems*. Ann Arbor: The University of Michigan Press, 1975.
6. P. Husbands, G. Jeremy, M. Ilhagga, and R. Ives. Two Applications of Genetic Algorithms to Component Design. In Terry C. Fogarty, editor, *Selected Papers: AISB Workshop on Evolutionary Computing, Lecture Notes in Computer Science No 1143*, pages 50–61. Springer Verlag, 1996.
7. R. Smith. A First Investigation into a Voxel Based Shape Representation . Technical report, Manufacturing Planning Group, Department of Mechanical Engineering, University of Edinburgh, 1995.
8. H. Watabe and N. Okino. A Study on Genetic Shape Design. In Stephanie Forrest, editor, *Proceedings of the Fifth International Conference on Genetic Algorithms*, pages 445–450. San Mateo: Morgan Kaufmann, 1993.

Phase Transition Networks: A Modelling Technique Supporting the Evolution of Autonomous Agents' Tactical and Operational Activities

Anthony G. Deakin and Derek F. Yates

Department of Computer Science, The University of Liverpool
Liverpool L69 7ZF
anthony@csc.liv.ac.uk, yatesdf@csc.liv.ac.uk

Abstract. The purpose of this paper is to introduce a modelling technique which the authors are using to evolve autonomous agents' action plans by means of genetic programming operations. The technique is described and its application is illustrated through examples. A brief outline is given of a formal model construction methodology that has been developed to accompany the modelling technique. Finally, the features of the technique are reviewed. Particular note is made of its suitability for modelling a broad variety of artificial and natural systems for problem-solving and domain exploration by means of evolutionary computation.

1 Introduction

The initial motivation for this paper was the requirement to find a representation to support the development of novel tactical and operational techniques for solving a given problem using genetic processes. The problem itself is based in the military field but the modelling technique used can be applied to a wide range of problems in differing domains. Military scenarios involve a potentially large number of autonomous and semi-autonomous agents engaged in various activities whose outcomes can interactively affect a number of protagonists (with plans that generally conflict). The scenarios require both reactive and proactive actions - the agents' activities are 'data-driven' and 'goal-led'. A feature of such scenarios is that the prevailing conditions in the scenario change as a result of the agents' activities (and as a result of external environmental changes). The agents have information gathering and processing facilities to monitor these changes and plans of action to deploy resources in achieving their goals subject to the conditions.

When modelling a 'system' or scenario in which autonomous entities are involved, there is generally a need to represent two major aspects: (i) some representation of the system components (entities) and the relations between them and (ii) some specification of what the entities can do - their actions and interactions, such as 'turn right', 'increase speed' or 'communicate with a neighbour'. The two aspects can be loosely termed the entity and the activity, declarative and procedural knowledge, or the noun and the verb, etc. Modelling both aspects at the same time can be problematic. A representation is required that can depict static relations as well as dynamic changes in the entity *over time*, effectively a dynamic model of a system's structure and functions. More complex still is the explicit, combined modelling of autonomous agents' purposes, capabilities and decision-making processes, in effect their cognition and action plans, in forms suitable for human understanding. In terms of human agents, for instance, Peterson in [7] views cognition as a process, in which the forms of representation themselves play a vital role. This underlines the problem of understanding models of cognition in order to, for example, apply them consistently. A third component is also normally required in order to instantiate the model with details of the constraints that govern how, when and what actions can be enacted; for example speed may be in the range 0 - 20 KM/H on a surface with incline less than 10 degrees, and so on. As well as the sufficiency of the representation for modelling these components, a fourth requirement is computational tractability and a fifth requirement is clarity and heuristic adequacy, in that the model should explain, or at least clearly express, the entity's decision-making (see Bench-Capon in [1]). This suggests that a representation be used for modelling the cognition and action of autonomous agents that is sufficiently close to natural human understanding.

In section 2, one modelling formalism, state transition networks, that might appear to meet the requirements for modelling complex autonomous agents, is considered. It is however perceived to have some deficiencies for this kind of modelling. An alternative formalism, phase transfer networks, is proposed. It is held that this formalism can overcome some of these deficiencies. In particular, the scaling up from a simple enumeration of states and transitions to a model of an agent's high level goals and strategies seems to require higher level constructs (phases). Section 3 introduces examples to illustrate how phase transfer networks can be built up starting from the level of one-dimensional state transition networks to ultimately take into account both agents' high level strategies to achieve goals as well as their lower level tactical and operational activities. In particular, it is suggested how the models can be useful for evolving agent strategy. Section 4 introduces the authors' methodology for constructing phase transfer networks by reference to the examples that have been presented. Section 5 discusses some salient features of phase transfer networks that appear to make them broadly suitable for modelling and evolving autonomous agents' plans. Section 6 summarizes the modelling approach and draws preliminary conclusions.

2 State Transition and Phase Transfer Networks

It would seem that state transition networks (STN) are adequate for representing autonomous agents. In the STN directed graph formalism, nodes are states and edges represent transfers between states. A system can be in any of a (possibly infinite) set of states defined by an n-tuple $\sigma = (\sigma_1, \dots , \sigma_n)$ of relevant variables. A simple example is the artificial ant as used for example by Koza in [5] where an input of 0 (representing the fact that there is no food ahead) results in an output action such as 'turn right' and a transfer from state 00 to state 01, whereas an input of 1 (food ahead) results in an action such as 'move ahead and eat the food' and transfer to state 00 again. Sloman advances a general theory of representation based on a conception of the intelligent agent as a control system with multiple control states (see [8] and [2]).

The main objections that can be raised against using STNs for modelling autonomous agents are, however, the complexity and consequent lack of clarity that arise when attempting to represent agents' decision-making. For even small sets of actions available to agents and of operators enabling them to ascertain state information, it appears necessary, when attempting to model the agents' capabilities, to construct a 'maximal' STN first. This consists in identifying all possible states, inputs and outputs and fully connecting the graph, which can be a computationally explosive process. It may or may not then be possible to prune some states and transitions. Human agents do not generally reason and act at this level of exhaustively identifying and considering options (cf. forward moves in chess). Rather, they have purpose, they form and adapt goals. For this reason, the authors use the concept 'Phase' so as to bundle states together, to mean a set of possible sequences of one or more states. When phases are instantiated, the sequence of states is possibly arbitrary. To assist with the bundling together of states, the authors have created a Methodology for the construction of a formal Phase transfer network System (M*Pha*Sys). The set of phases that is constructed represents the agent's strategy.

With the Phase Transition (or Transfer) Network (PTN), as the authors' version of the modelling technique is designated, and the MPhaSys methodology, it is believed all the requirements for modelling autonomous agents' plans can be met in a general (domain-independent) manner. The PTN consists simply of networked phases. Some phase is identified as the assumed starting point or state for the system to be in (may be arbitrary). A phase is similar to a state in the sense that an entity remains in a particular phase until something triggers a change to another phase. The 'something' that triggers a phase transfer consists, *at minimum*, of a state change, a single bit of information that can be observed and hence distinguished from other bits of information or noise. What the entity *does* while in a particular phase consists, *at minimum*, of a wait 'activity'. The value of the PTN is, however, that complete programs of an entity's goal-seeking activities rather than just single actions can be encapsulated into phases, thereby representing tactics over time. Secondly, that these programs may be generated by, for example, evolutionary computing techniques.

3 Examples of PTNs

Simple examples are initially presented to illustrate how PTNs may be used and how they are related to STNs. Subsequently, the MPhaSys methodology is briefly outlined and its use in developing more complex agents' plans representing both strategic and tactical information is discussed. To take an example of a man-made system rather than a natural system such as a comet travelling through the solar system, one can consider the case of a standard traffic light (UK version) which cycles (hopefully) between red, red and amber, green, amber and red again. (Many further conditions could apply such that lights change in accordance with traffic densities etc. These cases are not considered here and simplifying assumptions are made such that the actual timings of the lights are not relevant. There are also many non-meaningful states such as red and green simultaneously, or no lights, or flashing lights etc. These are considered later). As the PTN is highly suitable for graphical display, a diagram of the four-phase traffic light system is given in Figure 1.

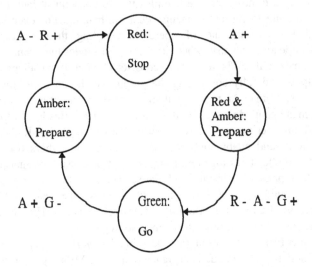

Figure 1. Four-phase lights PTN: the three lights R, A, G switch on (+) and off (-)

The lights PTN may be interpreted in terms of how a participant in a traffic system (an intelligent, autonomous agent, biological or artificial) should act when approaching the lights (assuming they are functioning properly), as follows:

> when the red light comes on - stop;
> when the amber light comes on - prepare to go;
> when the green light comes on - go;
> when the amber light comes on - prepare to stop.

The lights PTN can therefore describe the (intended) actions (reactions) of an entity (driver) within the scope of the lights-driver-road system. Thus far, the system can also be adequately modelled by some other representation such as a state transition table or decision table, e.g.:

	R	A	G
Stop	x	-	-
Prepare to go	x	x	-
Go	-	-	x
Prepare to stop	-	x	-

On reflection it may be seen, however, that the current lights PTN excludes the representation of an entire dimension where the driver is doing something in the meantime, such as starting, stopping, accelerating and decelerating. It assumes that an entity can react instantly to phase transfers, which is not the case where entities are moving, and thinking and braking distances are involved. However, a second PTN ring cycle may be 'interphased' with the first in such a way that the model now considers whether the entity is in the situation of having reached the lights or not (i.e. has come within stopping distance of the lights or not). The second lights PTN is depicted in Figure 2.

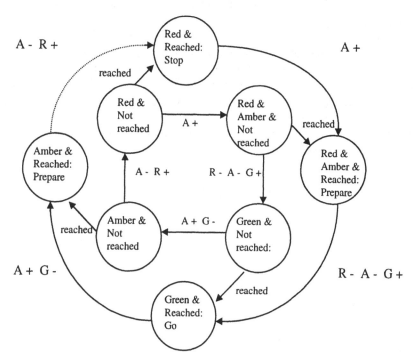

Figure 2. The dynamic (multi-phase) lights PTN: as long as the lights have not been reached, motion proceeds (the inner ring); when, however, the lights are within reach, transfer to the outer ring takes place. (The top left transfer is shown dotted to represent a possible 'grey' area in which the lights change to amber just as the entity has reached the lights but there is no time to stop - the entity might then 'clear' the lights).

The PTN is suitable for a 'one-off' approach to a set of lights. To obtain a further idea of how this situation can commonly apply in practice in an extended way, consider also a long straight road with lights visible in the distance that are changing quite frequently; it is not unusual to see complete light change cycles occurring, without, of course, any need to react to the lights *until* within reach of them. (As the default is generally for the lights to be green, the cut-off point of when the lights have been 'reached' is when the entity's safe stopping distance envelope - a function of speed - is within reach of the lights). The authors consider that the information contained in this second lights PTN is too complex to be portrayed meaningfully in decision table format for the purposes of conveying the nature of this system but it is clear in diagrammatic PTN form. Further, it will be noted that the phase transfer *circumstances* have been conjoined here - e.g. 'red **and** reached', 'red **and** not reached'. Also, that the PTN subsumes STNs. The current PTN is an STN in which phases are labelled states and the single *action* component that applies in a phase is also explicitly stated. This could be extended to more complex scenarios in the phases to make a subsuming PTN. For example, the entity might have a program of alternative action possibilities such as in the Green and Reached phase: 'if it is Monday and the traffic is heavy then turn right, else if there are no roadworks on the left turn left, else go straight on'.

As a second example of applying PTNs, a simplified version of the biological domain of Blood Glucose Regulation (BGR) ([4], [6]) is presented. Energy is obtained from food in glucose form but it is stored as glycogen in the liver. When energy is required by tissues such as muscle, the liver converts glycogen back to glucose and releases it into the blood. Hormones signal when the conversions are needed. The body ensures that the level of glucose in the blood is neither too high nor too low by means of a self-regulating, negative feedback process. Figure 3 illustrates use of the PTN, by way of example, to model this system.

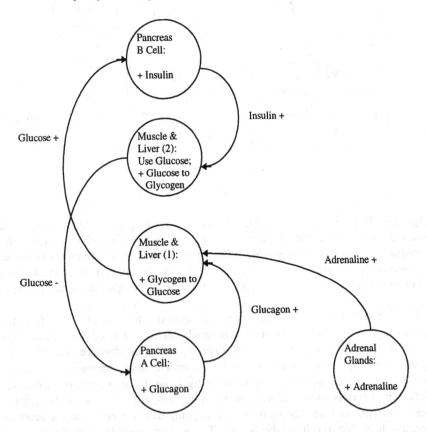

Figure 3. Simplified BGR: the directed arc label 'Glucose +' means '*when* glucose increases'; the action '+ Glycogen to glucose' means 'increase the conversion of glycogen to glucose'. Other inputs and outputs are treated as exogenous for the purposes of this example.

The PTN in Figure 3 has been drawn in such a way as to highlight the negative feedback loop by using a 'figure of eight' for the phase transfers between muscle/liver and pancreas A and B cells. The PTN may be interpreted such that when adrenaline increases and/or glucagon from the pancreas (A cells) increases, the liver (and muscles) convert glycogen to glucose, glucose levels rise, the pancreas (B cells) increase insulin levels, the muscles and liver use glucose and the liver converts surplus glucose back to glycogen for storage. The falling glucose level signals the pancreas (A cell) to release more glucagon again etc.

A third example concerns the patterns of action options available to an entity in a specific environment, an ecosystem, where its own actions and those of its neighbours interact. The neighbours may be using identical or different plans of action (and they could learn from each other). A very simple entity here might have the ability to ascertain whether a neighbour is N, S, E or W and to formulate actions from the set of go north, go east, go west, go south. Its circumstance-action pairings could then be for example (arbitrarily) 'if a neighbour is N - go S'; 'if a neighbour is S - go E, go N'; 'else go W; go N; wait' and so on. This simple plan is depicted in Figure 4. Time is incorporated in the scenario by means of extending the 'if' conditions to 'when' conditions.

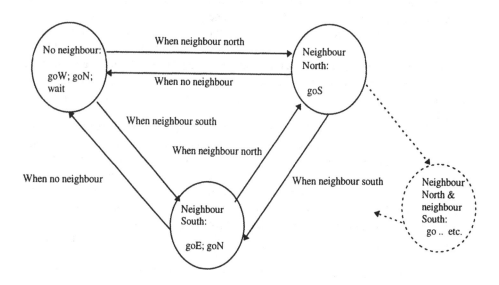

Figure 4. 'Ecosystem' PTN: the dotted node illustrates initial stages when adding a further phase.

Taking the scenario of Figure 4 a step further, it is possible to designate the particular phases as being sub-processes designed to achieve particular outcomes. It may be imagined that the scenario is one in which an animal is variously engaged in (the phases of) obtaining food, seeking or making shelter, rearing young, avoiding predators, and so on, autonomously and in cooperation with others. In distinction to the lights system where the goal is simply to proceed safely, and the BGR system, where the goal is to maintain regular glucose levels to meet requirements, there are a number of sub-goals and priorities in the ecosystem scenario. The resulting phase map can be likened to a tactical landscape (the phases) with lower level operations (the 'how' to achieve the 'what'). The autonomous agent has problems to solve or aims to achieve and the tactics it uses to solve or achieve them are represented in the phases. The arrangement of the phases constitutes a strategy.

The authors are currently working on a much more complex, scaled up version of Figure 4 with which plans of action for entities in various scenarios can be evolved using, for example, genetic programming (see Koza in [5], Deakin and Yates [3]). The PTN is used to model tactical and operational details that represent best practice for entities engaged in interactions. Genetic operations are then used to engineer better plans and solutions to problems.

4 The Formal PTN Methodology

As noted, the authors have created a Methodology for the construction of a formal Phase transfer network System (MPhaSys). The methodology proposes systematic steps to follow in constructing the network, involving identification of the (action-triggering) *circumstances* and the (circumstance-generating) *actions* available to entities and of the constraints that lead to their sequences. When a goal can be established for an entity, the appropriate PTN can be configured which will represent a plan of action to enable the goal to be accomplished. The methodology provides complete identification of meaningful phases/states; represents phases at the tactical and operational levels; and assists with ensuring consistency is maintained throughout the network. As devised, it permits genetic operations to be performed on the phases in such a way that an entity's set of phases (plan) can be attuned to the requirements of a scenario. (The authors describe this as 'cogenically' engineering plans - using computational genetics to evolve plans). Simple genetic crossover operators have been successfully applied and it is planned to introduce mutation operators as well.

In broad outline, the methodology proceeds as follows. A goal is identified, for an entity to, for example, find a specific neighbour in the ecosystem and then to navigate to another location (all without bumping into neighbours). The PTN is configured to represent the expected number of phases and transfers that are required to achieve this given the information and action resources (sets of circumstances and actions) available to the entity. Genetic programming is then applied to a population of PTNs and a plan to achieve the goal is evolved from the programs in the phases. Of particular interest is the reconfigurability of PTNs to represent agents' strategies for accomplishing new goals.

The methodology is illustrated by reference to the examples that have been introduced. In terms of the Lights PTN, the goal might be for an entity to pass the set of lights successfully. This is signalled by arriving in the 'Green and Reached' phase and then performing the action(s) associated with that phase - go (proceed), turn left, etc., as appropriate. The goal is thus 'Green and Reached'. This phase could then be shown as a terminating phase, without any out-going transfers. In order to get to this phase, a number of preceding transfers are required, which in turn determine predecessor phases and so on. (In the examples, the starting phase is arbitrary and the PTNs are cyclical). By backward chaining from the goal state via actions and transfer circumstances, it is possible to arrive at a PTN configuration consisting of at least one chain of circumstance-action pairs from initial to goal state. The pairing suggests the maximum number of phases that are required (as a flat STN). Subsequent PTN optimizations (using for example genetic operations) can however result in a phase not being used (e.g. it could contain only a 'wait' activity). Alternatively, a complex set of tactics might be developed in a single phase.

In the blood glucose regulation example, the goal of this sub-system is in fact to maintain adequate glucose levels. The system goal is an oscillation between two sub-goals of increasing and decreasing the glucose level according to the particular parameter settings (which have not been described). The particular phases and transfers depicted arise from the set of actions (e.g. increase glucagon production) and the set of circumstances (e.g. when the blood glucose level increases). This system represents however a case in point where the PTN could be decomposed into many lower levels of shorter time-slice PTNs subsuming STNs.

The ecosystem PTN example looks like an STN in which all states are made explicit, but which has the beginnings of complex programs of action (e.g. goW; goN; wait) embedded in the phases. The MPhaSys methodology would be applied here to again identify the expected phase (strategy) sub-map to achieve a goal state from an initial state based on the sets of information gathering operators (circumstances) and actions relevant to that scenario. Optimizations of the tactics (e.g. goW; goN; wait) with evolutionary computation methods are again germane to this example.

5 Features of the PTN

Although formal comparison with other modelling techniques is incomplete at this stage, it is apposite to note some of the characteristics (that emerge also from applying the PTN in the above examples) that contribute to it being domain-independent.

Formality, Semantic closure and Abstraction

A methodology is available, as noted, to enable PTNs to be formally specified and the authors are currently using MPhaSys to evolve plans for entities by applying Genetic Programming to PTNs. The methodology uses maximal identification of the components to be modelled. Significantly, however, it is not always necessary or even always possible to identify all components, such as states. With traffic lights, for instance, there are innumerable possible sequences when intermittent faults arise. These may be treated as faults 'by default', however, since they would over-ride normal sequences. Just one error condition is thus required even though there are countless ways in which it may arise. In other words, it is not necessary to exhaustively enumerate and model all contingencies - just to abstract those of interest. Doing this still enables the consistency and completeness of the PTN to be checked.

One is not really concerned how the lights change as long as an observer can distinguish between them e.g. between red and green (i.e. they are 'sufficiently discrete'; there is an appropriate time interval for the change such that they are effectively not fading into each other and are not both off).

Levels of Abstraction

As seen in the case of the blood glucose regulation and the ecosystem PTNs, the models are initially described at a level which is 'uninstantiated' - parameters have not been described. The circumstances and actions may be quantified to arbitrary complexity, for example - 'when glucose reaches x %, release y mg insulin'; 'when the neighbour is 5 miles north wait 10 minutes' and so on. Genetic operators such as crossover and mutation can then be applied to the parameters as well as to the operations and transfers.

Levels of Complexity, Scale and Intelligence

The phases may be arbitrarily complex, ranging at minimum from 1 action per circumstance to phases which contain entire programs with their own control structures.

The entities modelled may range from atomic components to large scale intelligent agents. With the traffic lights example, the lights system has 0 or more intelligent agents and is intended to apply to them all equally. In the glucose example, the entities are tissues.

PTNs can be assembled together into larger systems, for example, in the case of the lights, into coordinated traffic systems, or in the case of autonomous agents, into ecosystems with cooperation and competition where plans differ across agents. They may represent natural and man-made systems.

Stretching time

Time may be incorporated in the PTN in such a way that it may or may not have operational duration (for example, in simple sequences and in phase transfers it is possible that elapsed time can be ignored).

Object Orientation

PTNs are naturally suited to represent systems in an object-oriented manner. A single phase can represent an object with its set of actions. Messages are passed to other objects as phase transfer statements. An entity can be represented as a set of lower level objects (phases).

6 Summary and Conclusions

The M*Pha*Sys methodology can be used to construct formal PTN computational models. It assists with completeness and consistency checking. The PTN models can be used to represent agents' plans and these may then be evolved through the application of evolutionary computing approaches such as genetic programming to phases (tactics and operations) and/or to phase transfers. The evolved models can be useful to solve problems in novel ways and to assist in understanding a domain. The PTN modelling technique appears to be suitably flexible and adaptable for modelling biological and man-made systems. The models can be informal, semi-formal or formal. Informal and semi-formal models can have useful explanatory function. The formal computational models are useful for problem-solving and exploratory simulation as well as for explanatory representation.

References

1. Bench-Capon, T. J. M.: Knowledge representation. San Diego, CA: Academic Press (1990)
2. Dartnall, T.: Review of Peterson, D. (Ed.): Forms of representation: an interdisciplinary theme for cognitive science. Intellect Books (1996) in AISB Quarterly **97** (1997) 8-9
3. Deakin, A. G., Yates, D. F.: Genetic programming tools available on the web: a first encounter. In GP-96, Koza, John R., Goldberg, David E., Fogel, David B., and Riolo, Rick L. (editors): Genetic Programming 1996: Proceedings of the First Annual Conference, July 28-31, 1996, Stanford University. Cambridge, MA: The MIT Press (1996) 420
4. Jones, D.: Blood glucose regulation. Honours project report. Department of Computer Science, the University of Liverpool (1992)
5. Koza, J. R.: Genetic programming: on the programming of computers by means of natural selection. Cambridge, MA: The MIT Press (1992)
6. Paton, R. C.: Understanding biosystem organisation - part 1: techniques. International Journal of Science Education **15** (1993)
7. Peterson, D.: Introduction. In Peterson, D. (Ed.): Forms of representation: an interdisciplinary theme for cognitive science. Intellect Books (1996)
8. Sloman, A.: Towards a general theory of representations. In Peterson, D. (Ed.): Forms of representation: an interdisciplinary theme for cognitive science. Intellect Books (1996)

AN EVOLUTIONARY, AGENT-ASSISTED STRATEGY FOR CONCEPTUAL DESIGN SPACE DECOMPOSITION

I C Parmee and M A Beck
Plymouth Engineering Design Centre
University of Plymouth
Drake Circus, Plymouth PL4 8AA
email: mbeck@plymouth.ac.uk   iparmee@plymouth.ac.uk

Abstract: Genetic algorithm based strategies for the rapid decomposition of complex, conceptual design spaces into discrete, bounded regions of high performance are described. The objective is not to identify single peaks but to identify high performance regions. Sufficient regional cover (in terms of number of solutions) is required for the extraction of information relating to design characteristics to support the designer in decision making processes for the definition of optimal design direction. Three strategies using single population and parallel GA implementations are described and results are presented. An adaptive filter is introduced to eliminate a need for apriori knowledge of the design space. Rule-based agents complement the search process by exploring identified regions to define region bounds.

1   INTRODUCTION

It is proposed that during conceptual design an exploratory tool is required which provides an efficient search of a high-dimensional space described by the variable parameters of the system under design. The objective of such an exploration is the identification of a number of high-performance design regions. Regional information can subsequently be extracted to provide an indication of relevant design characteristics in terms of, for instance, solution sensitivity [1], design preference, manufacturability, etc.

During the higher levels of design a breadth first design approach is generally evident. The emphasis is therefore upon exploration rather than exploitation. The research explores the utility of evolutionary search within this high level design environment by investigating the relative performance of both panmitic and parallel GA implementations for high-performance region identification upon a number of test functions. This complements associated work [2, 3] related to the identification of optimal design direction (in terms of least risk and best performance) from high-risk conceptual design environments.

Although this GA-based decomposition of a complex, conceptual design space relies upon quantitative criteria, the analysis of solutions within such regions provides sufficient information for the engineer to form a qualitative judgement. This benefits the significant requirement for caution related to inherent uncertainty during the early stages of design . Further design effort can then be concentrated in appropriate high-potential areas that best satisfy possibly ill-defined criteria at least risk. The objective therefore is not to locate the global optimum of the space described by the mathematical model nor to cluster solutions on individual peaks. These approaches are not considered appropriate due to the poor definition of conceptual design models. Expending computational effort for the isolation of individual solutions when confidence in their validity is low is questionable. In this case regional identification could be considered a more valid approach.

2 INITIAL STRATEGIES

Initial research [1]. concentrated upon the establishment of simple variable mutation regimes that encourage diversity during the early stages of a GA search and promote the formation of clusters of high-performance solutions. Populations from selected generations are extracted and stored in a final clustering set. A near neighbour clustering algorithm [4] which requires little apriori knowledge of the design space, then identifies naturally occurring clusters from the stored data.

These cluster orientated GA strategies (COGA's) were integrated with a conceptual turbine blade, cooling hole model. The objective was to identify robust design regions that offer alternative design direction. The application to a six dimensional model of a gas turbine blade cooling channel resulted in the identification of discrete regions of high-performance in terms of minimum coolant flow rate. The relative sensitivity of each region was determined from a perturbation of solutions and a calculation of the standard deviation of the resulting coolant flow rates. However, the dimensionality of the problem made it impossible to identify characteristics of the process and to develop strategies for improvement. Validation of the results required exhaustive search of the identified regions and the surrounding areas. A series of experiments involving two dimensional test functions was therefore implemented in order to better understand the characteristics of the cluster-oriented approach [5] and these functions have been utilised throughout the programme of research in this area. The objectives of these experiments have been to:

- identify maximum number of regions
- improve set cover i.e. population density of each region
- improve the robustness of the techniques
- minimise number of test function evaluations

The initial work was inspired by a desire to include solution sensitivity (i.e. the sensitivity of the fitness of a design solution to mild perturbation of the variable parameters that describe it) as part of the overall fitness criteria. A perturbation-based sensitivity analysis of each solution is not feasible due to the number of required calls to the design model. However the identification of high-performance regions followed by the perturbation and subsequent sensitivity analysis of the design solutions within those regions allows the relative robustness of each region to be assessed at an acceptable level of computational expense. Further search can then be continued in those robust, high-performance regions that satisfy some preset sensitivity criteria.

The test function of Figure 1 provides an experimental basis for the establishment of the techniques with particular regard to the sensitivity issue. Two distinct regions are apparent. The reasoning is that solutions from region B may be preferable to those of region A where slight perturbation of the variables may lead to severe degradation of the design fitness. A compromise between design performance and design sensitivity must be made. Sensitivity however is only one example of design information that can be extracted from high potential regions.

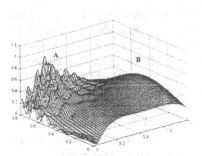

Figure 1: Surface plot of function F1

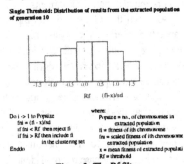

Figure 2: The Rf filter

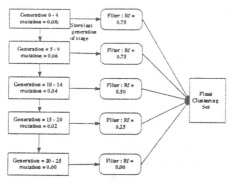

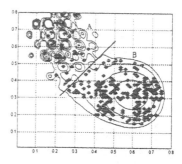

Figure 3: Varying mutation rates and filter rates

Figure 4 Contour plot of region derived from lower bound of Rf filter.

An adaptive filter has been introduced which prevents low performance solutions passing into the clustering set (i.e. the final set comprising of the extracted populations of pre-selected generations). The filter takes into account the relative fitness of the solutions from each of these populations by initially scaling them as shown in Figure 2 and introducing a threshold value (Rf). Solutions are either rejected or passed into the clustering set depending upon their scaled fitness in relation to the threshold value. The scaling allows the value of Rf to be preset or varied at each of the generations from which the populations are extracted and processed for inclusion in the final clustering set (see figure 3).

The Rf threshold is adaptive in that its value is relative to known solutions describing the surface topography at a particular time. The variation of Rf can be utilised in an investigatory manner during initial runs to investigate the relative nature of differing regions of the design space [5]. A sample output from the adaptive filter of function F1 is shown in figure 4.

Modifications and additional operators have been introduced to stimulate diversity and to address, in particular a lack of robustness of the strategy. Significant improvement has been achieved by introducing the following modifications to the basic algorithm:

- an initial structured population based upon a minimum fixed Hamming distance between each member of the population to ensure satisfactory cover of the search space at generation one[6].
- double-point reduced surrogate crossover to reduce incest and promote diversity [7]
- stochastic universal selection (SUS) [8]

Any members of the populations of those intermediate generations (i.e. those which do not contribute to the final clustering set) which have a fitness equal to or greater than that related to the Rf value of the adaptive filter are now stored and added to the final clustering set. This prevents the generation and subsequent loss of potentially useful information due to mutation. The following alternative structures have also been introduced

3 PARALLEL GA's

The parallelisation refereed to here involves dividing the single large population GA into a series of smaller sub-populations each of which evolve semi-independently with respect to the actions of crossover and mutation. Such independent evolution promotes speciation and the exploration of differing regions of the search space. Migration of genetic information from one species to another is an essential part of the parallel implementations.

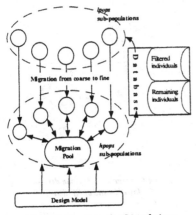

Figure 5: Topology of cgGA technique

3.1 COARSE-GRAINED GA ARCHITECTURE (cgGA)

The impetus for this model stems from two aims:

- To develop concepts from injection island GA's [after 9] concerning hierarchically arranged sub-populations and the notion of multi-level representation at differing resolution.
- To utilise a database of previous search results to identify under explored and possible feasible areas of the design space.

High-resolution (*hpops*), and low-resolution (*lpops*) subpopulations are introduced (see figure 5). Each of the *hpops* consists of twenty individuals with the fitness of each being determined by calls to the design model. As described in section 2, individuals are stored off-line for filtering. The remaining generations are also kept but in a separate database.

Migration between the *hpops* is based upon Euclidean distance: Excluding the fittest individuals, at the end of each generation five chromosomes are chosen randomly and placed into a 'migration pool'. The Euclidean distance between each individual in this pool and the fittest individual of each sub-population is determined and individuals migrate to the sub-population which contains the closest fittest individual, with the restriction that sub-population size remains at twenty chromosomes.

Individuals of the *lpops* concurrently search the database of extracted and filtered individuals. Each individual monitors a surrounding hypercube of pre-set dimension and that individual's 'score' is a function of the number of times its associated hypercube has been visited and the average fitness of those visits.. Thus, in general terms, those *lpops* individuals occupying a hypercube of high average fitness but a low frequency of visits receive a high overall score. Highest scoring individuals are then injected into the fine resolution sub-populations.

The topology between resolutions is one-to-one, i.e. each low resolution sub-population is related to a particular fine resolution sub-population to which individuals migrate copies of themselves. To make the transition from coarse to fine the least significant bits of the fine resolution are randomly generated (Figure 6) The twenty-individual populations within the fine resolution sub-populations are maintained either through an elitist strategy where the weakest is replaced or by replacing individuals randomly.

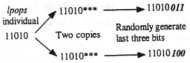

Figure 6: Transition of individuals from low to high

Hpops are run for n-initial generations which seeds the database and initiates regional search. This process is then paused and *lpops* are run for n generations at the end of which individuals migrate to *hpops* as described. A cyclic process then proceeds with *hpops* running for m generations followed by *lpops* for n generations (Figure 7). As the changing database, with each run of *hpops*, represents a dynamic fitness landscape for *lpops*, all *lpops* sub-populations are re-initialised at the beginning of each set of *m* generations.

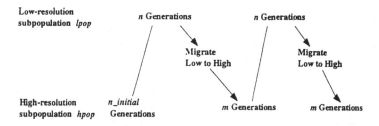

Low-resolution
subpopulation *lpop*

n Generations

n Generations

Migrate
Low to High

Migrate
Low to High

High-resolution
subpopulation *hpop*

n_initial
Generations

m Generations

m Generations

Figure 7: Interaction between the high and low resolutions

3.2 FINE-GRAINED GA ARCHITECTURE (fgGA)

The technique here is to structure a large population into a series of smaller sub populations by placing one individual at each location on a NxN network wrapped at the edges to form a torus. By defining a sub-population as a location plus its eight neighbours, i.e. a 3x3 grid, a sub-population is randomly chosen from the network and evolved in a similar fashion to a steady-state GA [10]: Roulette wheel selection is utilised to select two individuals for mating and mutation. Each resulting offspring is then compared with a randomly selected individual from its parent sub-population and is either rejected or replaces that individual dependant upon relative fitness.

Using this topology many authors [i.e. 11, 12] have reported the establishment of relatively stable genetic diversity through the emergence of *demes*: Clusters of genetically similar individuals. Moreover a local search occurs within a cluster and new regions are explored through the recombination of individuals at the edges. Although the same authors also report that the clusters are not stable and one deme will finally dominating the whole network, research suggests that uniform crossover with a high mutation rate may slow the convergence. [12].

To further reduce convergence rate and enhance the exploratory nature of the fgGA technique, a mechanism based upon phenotypic similarity is introduced to identify when a sub-population is converging upon a particular area of the search space and to mark that area as 'tabu'. The mechanism is based upon the observation that as a deme converges the intra-deme distance between individuals decreases whereas the inter-deme distance increases. The same is also true of sub-populations that are wholly contained within a deme and those that overlap two, or more, demes, (Figure 8) Convergence of a sub-population can be measured as a function of some ratio λ $(0 < \lambda < \lambda)$ of the distance between its individuals at initialisation to their current distance, i.e. dist $_{t+1} < \lambda*$dist$_0$.

Upon convergence the minimum and maximum limits of each dimension define the hypercube which then represents the initial definition of a region or partial region. This region is then considered 'tabu' with any new individual entering it being assigned a fitness of zero. The converged sub-population is then reinitialised with randomly generated individuals.

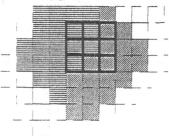

Figure 8: A 3x3 sub-population overlapping two demes

4 BOUNDARY IDENTIFICATION

It was noted in section 1. that the designer needs an exploratory tool to identify a number of high-performance regions and an indication of the design characteristics of those region. To address this a complementary rule-base/local search technique is introduced which:

- Enhances exploration - An increase in dimensionality greatly increases the probability of premature convergence upon a few high-performance regions resulting in non-identification of other regions of possible high potential. By utilising the minimum fitness given by the adaptive filter, the objective is to define the bounds of potential regions, based upon a hypercube defined by max and min parameter values for each dimension. Suppressing fitness within these roughly bounded regions allows further GA search to continue and other regions to be identified.

- Identifies those sub-regions that may exhibit different design characteristics, based upon a qualitative assessment of such characteristics.

It is not necessary to adopt a high resolution approach to boundary identification when preliminary design models are providing the relative fitness data. A computationally efficient strategy that identifies a few points is more realistic. A clustering algorithm is applied to the individuals which survive the adaptive filter and the outlying points of the resulting clusters are considered to represent approximations to the real boundary. The centre of the cluster is computed and initial search directions are established along lines from that central point running through each peripheral point of the cluster. Individual search agents lying on the initial bounds of the cluster then implement a simple line search to extend the boundary of the cluster relative to the lower bound defined by the adaptive filter.

Modelling both clusters and agents as objects, with each cluster having a set of agents as a slot value enables results from an agents search to be passed up to the parent cluster. This provides information upon which a qualitative assessment of the cluster's characteristics can be based. A globally set rule-base governs the agent step size and actions, such that:

i) Minimum step size, $S_{min.}$, is heuristically set equal to the resolution of the initial GA search.

ii) Each agent is allowed to 'leap' some multiple m, $\{m : m \in \{2^0 ... 2^n\}\}$ of $S_{min.}$. If this action brings the agent into contact with an obstacle, i.e. another cluster or fitness below the adaptive filter lower bound then the agent returns to its last position and m is halved. Each agent maintains its own value of m.

iii) By utilising a notion of *'crowding_space'* and *'empty_space'*, an agent violating another's *'crowding space'* is removed whereas if the *'empty space'* is underpopulated then agents are added at appropriate locations.

When $m = 1$, then obstacles take on another meaning:

i) Entering another cluster triggers a rule base which compares the characteristics of each cluster and if they are deemed similar then the clusters merge and all agents now belong to the merged cluster, otherwise the agent is stopped.

ii) An agent below the boundary threshold continues for three steps and if it renters an area above threshold it carries on searching as before. Otherwise it returns to the position where it first fell below threshold and stops. This gives an agent some momentum which helps overcome problems associated with non-convexity.

Qualitative assessment of each region is based upon the distribution of identified fitness values expressed as a probability density function from which the moments provide the necessary information, i.e. standard deviation is used to assess design sensitivity from a perturbation analysis of the identified solutions. Averaging sensitivity across a region gives an indication of the sensitivity of the surface. Skewness measured relative to either region mean or the mean across all regions assesses the symmetry of the fitness distribution around the chosen mean. Finally, kurtstosis can indicate the peakedness of the fitness distribution. All of these low-level measures can be easily computed from a perturbation of the results from the GA and local search processes. However their quantitative nature does not lend itself to qualitative assessment, which is seen as an essential requirement when considering the differing criteria upon which region similarity is to be based. Fuzzy logic is introduced to capture the continuity inherent in concepts such as design sensitivity, and an appropriately structured rule-base is able to assess the cumulative effects. Using this qualitative approach regions can be isolated by the extent they differ in relation to the degree of design sensitivity, i.e. standard deviation, and the degree of skewness of their fitness distributions. A more in-depth description of this qualitative assessment of regions can be found in [13].

5   COMPARISON OF THE STRATEGIES

To evaluate the performance of the COGA, cgGA and fgGA strategies three test functions have been developed with varying dimensionality: two 2-dimensional functions F2 and F2a (see Figure 9). All fitness landscapes described by the functions contain many sub-optima outside of the specified regions which are themselves dispersed across the design space. The two main objectives of the testing are to investigate the:

i.   relative performance of the strategies in terms of region identification
ii.   extent to which each of the strategies are able to maintain concurrent search across several
      regions.

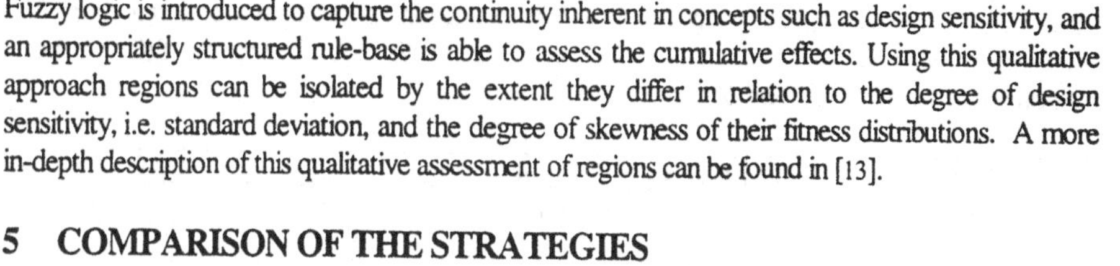

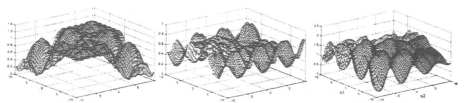

Figure 9: Showing surface plots of function F2, left, function F2a. middle, and a slice of function F4, right

To assess the exploratory nature of each technique the proportion of individuals within each region is compared against that expected. Expected region population proportion is estimated by exhaustively iterating across each region summing fitness exceeding the pre-defined lower bound of that test function. The method of calculating both expected and actual region population proportion is shown in Figure 10 Regions are defined a priori in order to establish a control for subsequent testing.

Expected population proportion
of region i

$$\text{Prop}(R_i)_{expected} = \frac{\Sigma f_i}{\Sigma\Sigma f_{i,j}}$$

Σf_i = sum of fitness across region
$\Sigma\Sigma f_{i,j}$ = sum of fitness across all

Actual population proportion
of region i

$$\text{Prop}(R_i)_{actual} = \frac{n_i}{N_{tot}}$$

n_i = no., of individuals in region i
N_{tot} = no., of individuals across all

Figure 10: calculating expected and actual region population proportion

The parameter settings for all strategies are shown in table 1 and twenty independent trials are run for each strategy. While the resolution of the search, see table 1, is high for an ill-defined preliminary design model, engineering design problems tend to be highly dimensional and a high resolution is adopted solely to compensate for the low dimensional test functions and to provide a more rigorous test of the strategies. For the 2D functions performance is monitored at 1500, 2500 and 3500 function evaluations. For the 4D function this is increased to 3500, 4500 and 5500 evaluations. With respect to the adaptive filter the populations are stored at five stages of the run and these stages are scaled appropriately i.e. for 3500 evaluations populations are stored every 700 evaluations with relative Rf values of 0.75, 0.75, 0.5, 0.25, 0.00. Testing the fgGA includes assessment of 16x16; 20x20 and 24x24 network sizes

Parameter	COGA	cgGA		fgGA
		lpops	hpops	
Selection type	SUS[1]	RWS[2]	RWS[2]	RWS[2]
Cross-over type	Dbl pt RS[3]	Dbl pt RS[3]	double point	uniform
P(Cross-over)	0.7	0.7	0.7	0.7
P(mutation)	see section 2.	see section 2.	0.01	0.025
Resolution	(4)	5 bits per variable	(4)	(4)
No., initial generations	-	-	5 generations	-
Run coarse every	-	-	5 generations	-
grid size	-	-	-	16x16

[1] Stochastic Universal Selection
[2] Roulette wheel Selection
[3] Double Point reduced Surrogate
[4] 10 bits per variable for 2D test problems, 8 bits per variable for 4D test function

Table 1: Parameter settings of techniques used in study

6   RESULTS AND DISCUSSION

Results for both F2 and F2a, shown in table 2, indicate that all strategies identify all regions within 1500 evaluations. The fgGA achieves the best set cover with the total population at 1500 evaluations similar to that of COGA at 3500 evaluations However a more apparent finding with these functions is the differential extent to which the strategies are able to populate regions with respect to the fitness

GA type	Evals		R1	R2	R3	R4	R5	R6	R7	TOTAL	R1	R2	R3	R4	R5	R6	R7	TOTAL
						FUNCTION F2								FUNCTION F2a				
						Region Number								Region Number				
COGA	1500	μ	6.9	37.8	7.6	2.0	5.9	32.2	7.7	99.9	9.8	6.9	5.7	192.9	41.8	36.5	16.3	309.8
		σ	5.5	18.2	6.0	2.0	4.9	17.8	4.5	58.8	5.0	4.3	3.7	39.8	15.2	19.4	10.7	98.0
	2500	μ	13.5	58.3	11.4	3.2	10.2	57.8	10.5	164.7	17.2	8.4	8.9	345.2	97.1	69.7	32.1	578.4
		σ	5.3	24.4	5.6	2.0	4.9	21.4	4.3	67.9	8.3	5.2	4.8	64.4	28.7	29.6	13.6	154.7
	3500	μ	15.4	100.6	17.7	4.5	13.5	102.0	20.7	274.3	26.5	15.5	11.5	562.2	153.5	93.7	92.7	955.6
		σ	10.2	87.8	12.2	2.8	8.1	65.8	10.0	197.0	8.1	4.7	5.5	64.4	40.6	42.9	15.8	182.3
cgGA	1500	μ	12.8	66.7	13.0	2.5	10.1	65.9	10.6	181.6	15.3	16.4	15.9	241.4	52.0	43.5	30.6	415.0
		σ	11.4	16.8	8.4	2.3	6.5	23.3	7.9	76.6	8.2	6.9	8.5	40.5	25.1	15.2	18.0	122.5
	2500	μ	15.3	113.3	18.6	4.7	15.0	117.3	17.3	301.4	22.7	20.6	22.3	402.7	72.2	76.2	59.0	675.6
		σ	6.7	34.7	11.1	3.8	10.4	38.2	9.6	114.5	7.8	8.5	13.5	54.0	17.5	23.4	17.6	142.4
	3500	μ	23.3	145.4	24.7	4.7	16.3	149.1	22.6	385.8	28.5	27.1	31.5	503.0	104.7	96.8	92.7	884.1
		σ	12.0	43.5	9.5	3.9	7.9	57.6	10.3	144.8	8.5	11.9	10.6	46.3	23.0	31.9	27.6	160.6
fgGA	1500	μ	11.9	118.8	10.1	1.9	9.8	99.3	9.4	261.0	12.6	10.5	8.3	319.3	86.1	59.1	33.8	529.5
		σ	5.5	41.5	6.5	1.8	7.8	37.9	4.5	105.5	7.4	8.6	5.8	45.2	34.4	22.7	15.6	139.7
	2500	μ	11.8	224.8	12.5	1.9	12.2	205.9	12.9	481.8	14.6	7.7	9.2	542.5	165.4	129.8	57.5	926.6
		σ	6.4	82.1	9.6	1.5	8.2	79.8	7.7	195.4	8.6	4.3	6.0	92.8	58.7	58.0	33.1	261.5
	3500	μ	32.5	392.2	33.9	3.7	23.5	387.5	27.0	900.1	15.2	10.4	8.8	764.2	251.8	170.9	79.7	1300.8
		σ	13.1	106.2	19.6	2.4	10.5	100.4	14.2	266.4	7.5	8.6	4.7	146.9	81.3	98.5	44.8	392.3
Expected Prop			0.07	0.36	0.07	0.02	0.06	0.35	0.07		0.05	0.04	0.05	0.59	0.09	0.09	0.09	

Table2: Number of individuals located within each region of functions F2, left table, and F2a, right table, average across twenty trials

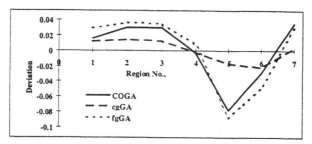

Figure 11: Function 2a - mean deviation of actual proportion .v. expected proportion

within those regions. The last row of table 2 gives the expected population proportion based upon relative mean region fitness. Figure 11 shows the observed mean deviation of actual region population proportion of F2a for each strategy at 2500 evaluations with respect to the expected proportion (positive values indicate over-population, negative values indicate under population). This shows that the cgGA achieves a closer correlation to the ideal of zero deviation.

Moving on to the 4D test function. Table 3 shows condensed results relating to mean total region population. Whilst all techniques are exhibiting a similar patterning to that of the 2D test functions, the performance of the fgGA deteriorates with increased network size.

GA Type			COGA	cgGA	fgGA 16x16	fgGA 20x20	fgGA 24x24
Evaluations	3500	μ	57.5	137.0	122.3	55.2	31.9
		σ	31.3	48.5	27.8	24.1	18.8
	4500	μ	74.7	199.0	206.1	111.6	67.4
		σ	20.9	101.1	69.2	38.4	34.7
	5500	μ	128.9	254.4	326.7	181.9	94.3
		σ	29.1	136.3	74.5	60.0	26.1

Table 3: Results summary for function F4

Individual region population is addressed in figure 12 which show the expected proportion and actual proportion for each strategy at 4500 evaluations using a 16x16 grid for the fgGA. Although all techniques are populating most regions, the COGA shows a tendency to favour one region whereas a closer match to the expected is achieved by cgGA and fgGA. However across all trials the same regions were not consistently identified. Figure 13 shows the standard deviation in observed proportional regional population, these indicate a relatively stable performance for COGA when compared to the erratic behaviour of cgGA. The performance of the fgGA lies between the two. Whilst it is not possible to discuss the results in depth a few general points can be made.

The deterioration in performance of the cgGA in F4 is attributable to the increasing dimensions: Approximating an area of previous search results is plausible only if there is information to average. With search space size governed by mn (m = bit resolution, n = no., dimensions), as n increases the probability of one individual being located in any one area is inversely proportional to mn. Hpops has far less chance of finding sufficient 'previous visits' from which the fitness function may give a reliable result.

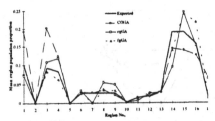

Figure 12: Function F4 -mean region population for
each technique and expected population proportion

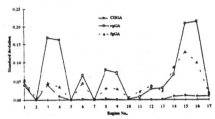

Figure 13: Function F4 - standard deviation in region
proportional population

The fgGA is exhibiting a certain degree of convergence on function F2a which diminishes on the 4D function. It is suggested that with lower dimensional problems, with fewer schemata to process and fewer regions to populate the fgGA's ability to populate regions has simply resulted in a particular state of convergence earlier.

To assess the effectiveness of the agent strategies for boundary identification the technique was tested on F1 and F2. For F1 the minimum step size was set at 8 bit resolution and the multiple leap distance m set at 16 thus giving a maximum resolution of 4 bits. An initial COGA search resulted in the identification of twenty clusters containing a total population of 524. Many of these were not at the cluster boundaries and therefore did not participate in the boundary search. Some 834 further evaluations were required to establish the three regions shown in figure 14. Although three separate regions have been identified the definition of the boundary between regions 1 & 2 is disappointing. Applying the strategy to F2 results in a higher definition of region bounds due to their more isolated nature (figure 15). Note, however, the inconsistencies between regions 8-10 and regions 2 and 3 in areas of the design space that are identical in surface characteristics.

The strategy is successful to some extent in that the rule-based agents are able to identify the regional bounds and the application of fuzzy logic to the isolation of sub-regions appears to represent a valuable tool. However the inconsistencies indicate a requirement for finer tuning of both the rule-base and the overall strategy. In particular the addition of qualitative design criteria in a real-world application may lead to a more succinct definition.

It has not been possible here to include all relevant results and associated discussion related to the research due to space restrictions. A more in-depth treatment can be found in reference [13].

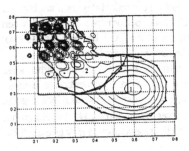

Figure 14: Results of boundary identification
for function F1

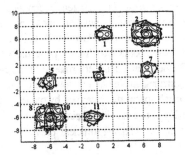

Figure 15: Results of boundary
identification of function F2

CONCLUSION

The paper assesses the application of both single population and parallel GAs to a novel area of research relating to automatic problem decomposition within the domain of conceptual/ preliminary design. The objectives of such research tend to differ from those of global optimisation in that the exploratory capabilities of the various strategies are, at least, of equal importance as their exploitation characteristics. Although the initial research commenced in a real-world domain the problems associated with high-dimensional visualisation of the strategic characteristics has resulted in experimentation involving low dimensional test functions.

Conclusions relating to the three tested strategies suggest that although relatively easy to implement and comparatively effective in evolving an initial diverse population, the COGA requires additional mechanisms to ensure that both inter and intra cluster mating occurs within and between emerging clusters. This will allow stability within a cluster and exploration of external areas.

The cgGA biases the search towards exploration by the injection of individuals between low and high resolution search procedures. The utilisation of a database of search results to guide the whole search process is a feasible strategy but further research is essential to address the appropriate processing of such information and the development of an injection hierarchy that does not result in overall search destabilisation.

The fgGA offers a less structured approach than that of the cgGA, relying upon asynchronous swapping of individuals between subgroups and a natural emergence of demes. Problems of deme dominance have been addressed resulting in a slowing of the process but further work is needed to develop a more effective mechanism to further promote concurrent exploration.

The rule-based search agents contribute significantly to the processing of the over-clustering approach applied to the GA search results. This leads to cluster merging and the eventual definition of region bounds. The use of fuzzy logic here is valuable but more qualitative information is required to better define regional characteristics. In a real-world scenario this qualitative information would relate to specific design requirement as described in associated research [14]. Unfortunately such information is unavailable from test functions. The conclusion at this time is, therefore, that the agent concept has been proven to sufficient extent to provide a basis for boundary identification as we return to real-world application.

The utility of the research described here must be assessed in terms of integration of evolutionary and adaptive computing with the engineering design process as a whole. The overall objective of this and related research [15] is the development of underlying search engines that can support the designer during conceptual, embodiment and detailed design. The automatic decomposition of large, preliminary design spaces into regions of high-performance/potential and the extraction of relevant information from such regions is a major objective. The work described in the paper is a significant contribution to the realisation of that objective.

ACKNOWLEDGEMENTS

The Plymouth Engineering Design Centre is one of six UK EPSRC Engineering Design Centres. This research has also been supported by Rolls Royce plc. We wish to thank these organisations for their continuing support.

REFERENCES

1   Parmee I. C., Denham M. J. (1994) The Integration of Adaptive Search Techniques with Current Engineering Design Practice. Procs. of Adaptive Computing in Engineering Design and Control; University of Plymouth, UK; Sept. 1994; pp 1-13

2   Parmee, I. C. (1997) Evolutionary and Adaptive Strategies for Engineering Design - an Overall Framework. *IEEE International Conference on Evolutionary Computation*, April 13-16, 1997.

3   Parmee, I. C, (1996) Towards an Optimal Engineering Design Process Using Adaptive Search Strategies. *Journal of Engineering Design* 7(4). pp 341-362.

4   Jarvis R. A. and Patrick, E. A. (1993) Clustering using a Similarity Measure Based on Shared Near Neighbours. *IEEE Transactions on Computers*, (22)11.

5   Parmee. I. C. (1996) Cluster-Oriented Genetic Algorithms (COGA's) for the Identification of High-Performance Regions of Design Space. *Procs. EvCA Conference Moscow.* June 24-27 1996

6   Reeves, C. R. (1993) Using Genetic Algorithms with Small Populations. *Procs. of the Fifth International Conference on Genetic Algorithms.* pp 92-99

7   Brooker, L. (1987) Improving Search in Genetic Algorithms, In Davis, L. (Ed) *Genetic Algorithms and Simulated Annealing* pp 61- 73. Morgan Kaufmann

8   Baker, J. E., (1987) Reducing Bias and Inefficiency in the Selection Algorithm. *Procs of the Second International Conference on Genetic Algorithms*, pp14-21

9   Lin, S. C. and Goodman, E. (1994) Coarse-grain parallel genetic algorithms: Categorisation new approach. *Sixth IEEE symposium on Parallel and Distributed Processing*, Los Alamitos, CA: IEEE Computer Society Press

10 Syswerda, G. (1989) A study of reproduction in generational and steady state genetic algorithms. In G. Rawlings (Ed.), *Foundation of Genetic Algorithms*, Morgan Kaufman.

11 Davidor, Y., Yamada, T. & Nakano, R. (1993) The ECOlogical framework II: Improving GA performance at virtually zero cost. *Procs of the Fifth International Conference on genetic Algorithms.* pp 171 - 175

12 Spiessens, P & Manderick, B. (1991) A Massively Parallel Genetic Algorithm: Implementation and First Analysis. *Procs. of the Fourth International Conference on Genetic Algorithms.* pp 257 - 263

13 Beck, M. A. (1996) Parallel Genetic Algorithms: An Investigation of Coarse and Fine Grained approaches to Design Space Decomposition. MSc Thesis. University of Plymouth.

14 Roy R., Parmee, I. C., Purchase, G. (1996) Integrating the Genetic Algorithm with the Preliminary Design of Gas Turbine Blade Cooling Systems. *Procs. of Adaptive Computing in Engineering Design and Control*; University of Plymouth, UK; March. 1996.

15 Parmee. I. C. (1996) The Development of a Dual-Agent Strategy for Efficient Search Across Whole System Engineering Design Hierarchies. *Parallel Problem Solving From Nature, Lecture Notes in Computer Science*, September 22-27 1996

Task Scheduling with Use of Classifier Systems

Franciszek Seredyński

** Institute of Computer Science, Polish Academy of Sciences
Ordona 21, 01-237 Warsaw, Poland
sered@ipipan.waw.pl

Abstract. A new approach to develop parallel and distributed algorithms for scheduling tasks in parallel computers with use of learning machines is proposed. Coevolutionary multi-agent systems with game theoretical model of interaction between agents serve as a theoretical framework for the approach. Genetic-algorithms based learning machines called classifier systems are used as players in a game. Experimental study of such a system shows its self-organizing features and the ability of emergent behavior. Following this approach a parallel and distributed scheduler is described. Results of the experimental study of the scheduler are presented.

1 Introduction

The problem of *scheduling* [4, 9] tasks of a parallel program into processors of a parallel system belongs to key issues influencing the performance of currently used or designed parallel computers. The problem is known to be NP-complete in its general form. Corresponding to them functions of operating systems are therefore rarely fully implemented. It decreases the efficiency of use of today's parallel computers and makes them difficult to use.

Over the years, a number of methods have been applied to attack the scheduling problem. Some insight into currently used methods of scheduling can be found in the recent literature [1, 18]. Most of these methods are able to find an optimal solution only in several restricted cases. To deal with the most general case of the problem, a number of *heuristics* have been introduced. They do not guarantee an optimal solution to the problem, but they find near-optimal solutions most of the time.

Most of known heuristics are *serial* algorithms which are executed on sequential machines. A new perspective line of the research in the area is developing *parallel* algorithms of scheduling [2]. Constructing parallel scheduling algorithms results in decreasing the complexity of heuristics and increases speed-up of the algorithms.

In the paper a new approach to develop *parallel and distributed* algorithms of scheduling is proposed. To develop such algorithms a theoretical framework

---

** also from Warsaw University of Technology, Institute of Control and Computation Engineering, Poland

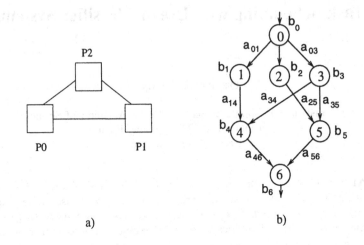

Fig. 1. Examples of a system graph (a), a precedence task graph (b)

based on a model of N-person *games with limited interaction* is considered. Genetic algorithms-based learning systems called *classifier systems* are used as players in the game. Such a system has the ability of a global behavior while only local interaction between players is observed. The game-theoretic model is used to build a parallel and distributed scheduler.

The remainder of the paper is organised as follows. The next section discusses accepted models of a parallel program and a parallel system in the context of a scheduling problem. Section 3 describes a game-theoretic framework and presents results of an experimental study of the coevolutionary multi-agent system with use of classifier systems. Section 4 contains a description of a proposed scheduling algorithm and details of actually implemented version of the scheduler. Results of experimental study of the scheduler are presented in Section 5. Last section contains conclusions.

2 The Scheduling Problem

Models of parallel computers and parallel programs which we accept are oriented on MIMD machines and are formulated as follows. A multiprocessor system is represented by an undirected unweighted graph $G_s = (V_s, E_s)$ called a *system graph*. V_s is the set of N_s nodes of the system graph representing processors with their local memories of a parallel computer of MIMD architecture. E_s is the set of edges representing bidirectional channels between processors and define a topology of the multiprocessor system. Fig. 1a presents an example of a system graph representing a multiprocessor system consisting on processors *P0*, *P1* and *P2* arranged in a ring topology.

If u and v are two vertices in G_s, (u, v) will denote the arc from u to v. The *distance* $d(u, v)$ between u and v is the length of the shortest path in G_s,

Table 1. Payoff function of a game on a ring

	$s_{k\ominus 1}$	s_k	$s_{k\oplus 1}$	$u_k^1(s_{k\ominus 1}, s_k, s_{k\oplus 1})$
0	D	D	D	10
1	D	D	C	0
2	D	C	D	0
3	D	C	C	0
4	C	D	D	0
5	C	D	C	50
6	C	C	D	0
7	C	C	C	30

connecting u and v, and measured in a number of hops between u and v. Two processors corresponding to vertices u and v respectively are called *neighbour* processors if the distance $d(u, v) = 1$. It is assumed that all processors have the same computational power and a communication via the links does not consume any processor time.

A parallel program is represented by a weighted directed acyclic graph $G_p =<V_p, E_p >$, called a *precedence task graph* (see, Fig. 1b. V_p is the set of N_p nodes of the graph representing elementary tasks, which are indivisible computational units. Weights b_k of the nodes describe the processing time needed to execute a given task on any processor of a given multiprocessor system. There exists a precedence constraint relation between the tasks k and l in the precedence task graph if the output produced by task k has to be communicated to the task l. E_p is the set of edges of the precedence task graph describing the communication pattern (connectivity) between the tasks. Weights a_{kl} of the edges describe a communication time between pairs of tasks k and l, when they are located in neighbour processors. If the tasks k and l are located in processors corresponding to vertices u and v in G_s, than the communication delay between them will be defined as $a_{kl} * d(u, v)$.

The purpose of *scheduling* is to distribute the tasks among the processors in such a way that the precedence constraints are preserved, and the *response time* T (the total execution time) is minimised.

3 Games of Classifier Systems

3.1 Games with Limited Interaction

As a theoretical framework for our approach to develop parallel and distributed algorithms of scheduling a model of a game with limited interaction [14, 17] will be used. A game with a limited interaction is given by a set $N = \{0, 1, \ldots, N-1\}$ of N players, set S_k of actions for each player $k \in N$, and a payoff function u_k which depends only on its own action s_k and the actions of its n_k neighbours in the game.

The simplest game with limited interaction is a game on a ring. The payoff function of any player in the game on the ring depends on his actions and on the actions of his two neighbours $k \ominus 1$ and $k \oplus 1$, where $\ominus$ and $\oplus$ denotes subtraction and addition modulo N. Assuming that the set S_k of actions for each player is limited to two alternative actions **C**-cooperate, and **D**-defect, the payoff function has 8 entries and Table 1 shows an example of a payoff functions u_k^1 used in our study.

A widely known Prisoner's Dilemma [3, 10] game theory model can be considered as a special case of the presented model. The dilemma to cooperate or defect is expressed in the model by entries 7 and 5 respectively of the Table 1. In opposite to the Prisoner's Dilemma model, we admit in our model any values of the payoff function. Exchanging e.g. values of the payoff function corresponding to entries 7 and 5 (see, Table 1) results in removing the effect of the dilemma, while the model is still interesting from point of view of real applications.

The most widely used solution concept for noncooperative games, represented by the game with limited interaction, is a *Nash equilibrium point*. A Nash point is an N-tuple of actions, one for each player, such that anyone who deviates from it unilaterally cannot possibly improve his expected payoff. A Nash equilibrium point will define payoffs of all the players in the game. We are interested in some global measure of the payoff received by players. This measure can be e.g. the average payoff $\overline{u}(s)$ received by a player as a result of combined actions' $s = (s_0, s_1, \ldots, s_{N-1})$, i.e.

$$\overline{u}(s) = \left(\sum_{k=0}^{N-1} u_k(s) \right) / N, \tag{1}$$

and it will be our global criterion to evaluate the behavior of the players in the game. The question which arises immediately concerns the value of the function (1) in a Nash point. Unfortunately, this value can be very low.

Analysing all possible actions' combinations in the game and evaluating their prices, i.e. a value $\overline{u}(s)$, we can find actions' combinations characterized by a maximal price and we can call them *maximal price points*. Maximal price points are actions' combinations which maximise the global criterion (1), but they can be reached by players only if they are Nash points. A maximal price point usually is not a Nash point and the question which must be solved is how to convert a maximal price point into a Nash point.

To solve the problem we introduce a cooperation between players, who share their payoffs in a game. Generally, on the base of potentially introduced cooperation into a game, we consider the following schemes of the game:

- game **G**: *no cooperation* in the game, i.e. a payoff of a player k is defined by u_k
- game **G***: *local cooperation*- a player k shares his payoff with his neighbours in a game, i.e. his payoff is transformed as:

$$u_k \longrightarrow w_k = \frac{u_k + \sum_{l \in n_k} u_l}{\max_{l \in \mathbf{N}} n_k + 1}, \tag{2}$$

where u_l is a payoff of a neighbour l of a player k, n_k is the number of neighbours of the player k, and $\max_{l \in \mathbf{N}} n_k$ denotes a maximal number of neighbours in a game

- game $\mathbf{G^{**}}$: *global cooperation*-sharing a payoff received by a player k by all players participating in a game, i.e. his payoff u_k is transformed into a new payoff w_k in the following way:

$$u_k \longrightarrow w_k = \overline{u}(s). \qquad (3)$$

We are interested in a global behavior (in the sense of the criterion(1)) of a team of players taking part in iterated games. Presented in this Section model has been considered from positions of a collective behavior as a coevolutionary multi-agent system [14, 17]. In this paper we use genetic algorithms-based learning machines called *classifier systems* [5, 11]as players in the game model.

3.2 Games of Classifier Systems

Classifier systems constitute the most popular approach to genetic algorithms-based machine learning. Recently, they have been successfully applied [6, 7, 12] to solve real-life problems. A general framework for studying the behavior of a game-theoretic *multi-agent* system when each agent-player is a classifier system has been proposed [16] recently. A learning classifier system (CS) maintains a population of decision rules, called *classifiers*, evaluated by using them to generate actions and observing received rewards defined by a payoff function, and modified by periodically applying genetic algorithms (GA).

A classifier c is a condition-action pair

$$c = < \text{condition} > : < \text{action} >,$$

with the interpretation of the following decision rule: if a current observed state matches the **condition**, then execute the **action**. The action part of a classifier is an element of the set S_k of actions a player k in a game with limited interaction. The conditional part of a classifier of CS representing a player k contains his action and actions of his neighbours in the game, and additionally a *don't-care* symbol #.

A usefulness of a classifier c, applied in a given situation, is measured by its *strength str*. A real-valued strength of a classifier is estimated in terms of rewards obtained according to a payoff function of a player k, using by the player the given classifier to generate an action. Action selection is implemented by a competition mechanism, where the winner is a classifier with the highest strength.

To modify classifier strengths the simplified *credit assignment* algorithm [11] is used. The algorithm consists in subtracting a tax of the winning classifier from its strength, and then dividing equally the reward received after executing an action, among all classifiers matching the observed state.

A strength of a classifier has the same meaning as a *fitness function* of an individual in GA (see, e.g. [11]). Therefore, a standard GA with three basic

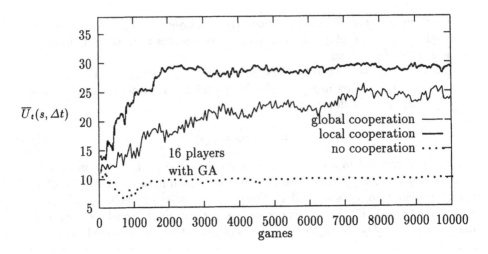

Fig. 2. Games of classifier systems without cooperation, with a local cooperation, and with a global cooperation

genetic operators: selection, crossover and mutation is applied to create new, better classifiers. Crossover is applied only to the conditional parts of classifiers. Mutation consists in altering with a small probability randomly selected condition elements or actions. GA is invoked periodically and each time it replaces some classifiers with new ones.

We will study below iterated games of CSs. An iterated game consists of a number T_g of single games $s(t) = (s_0(t), s_1(t), ..., s_{N-1}(t))$ played in subsequent moments of time $t = 0, 1, ..., T_g - 1$, with a value T_g unknown for players. In a single game played at the moment t each player autonomously selects an action to match an observed state $x_k(t)$ of a game environment. Observed by a player k state $x_k(t)$ of the environment is formed by his and his neighbours' actions played in a previous moment of time. Rewards defined by a payoff function are transferred to players directly or after their redistribution, if a cooperation between players exists.

A number of experiments with use of CSs as players in the iterated game on a ring has been conducted. A detailed discussion of CSs and GAs setting parameters used in experiments can be found in [16]. Fig. 2 shows some results of experiments with 16 players participating in an iterated game defined by a payoff function from Table 1. The figure shows the average payoff $\overline{U}_t(s, \Delta t)$ of players, received during each $\Delta t = 50$ games, i.e.

$$\overline{U}_t(s, \Delta t) = \left(\sum_{t'=\text{entier}(t/\Delta t)}^{\text{entier}(t/\Delta t)+\Delta t-1} \overline{u}(s, t') \right) / (\Delta t), \qquad (4)$$

in a game without cooperation, in a game with a local cooperation, and in a game with a global cooperation.

One can see that in the iterated game without a cooperation the multi-agent system converges to a steady-state with a corresponding the average payoff of a team of players equal to 10, defined by the Nash point. In the game with a global coperation a self-organization process can be observed, resulting in emerging a global behavior. The system converges to a steady-state corresponding to playing the maximal price game, providing the maximum value of the average payoff received by the players and equal to 30. One can see that in such a distributed system a global behavior evolves, and is achieved by an interaction between players, resulting in optimization of a global criterion.

The need of applying a specific model of a cooperation depends on a structure of a given application. If a given problem does not belong to a class of Prisoners' Dilemma model we can apply the model of a game without a cooperation. If a cost function of a given application can be decomposed into locally defined functions we can apply the model of the game with a local cooperation. When decomposition of the cost function is not known we apply the model of the game with a global cooperation.

4 Game Theoretic-Based Scheduler

To design a parallel and distributed scheduler we will interpret a parallel program represented by a precedence task graph as a multi-agent system with agents taking part in a game with limited interaction. We assume that a collection of agents is assigned to tasks of the program graph in a such way that one agent is assigned to one task. An agent P_k $(k = 1, 2, ..., N_p)$ has a number of actions a_k which influence a global function T describing the response time for a given scheduling problem.

If tasks of the program graph together with agents attached to them are placed in some way, e.g. randomly into the system graph, the agents' actions can be interpreted in terms of possible moves of the agents in the system graph.

Each agent is a CS learning machine and it is responsible for local mapping decisions concerning a given task. Each CS has a number of classifiers describing rules and corresponding to them decisions which are used in a process of an iterated game. Classifiers will describe accepted attributes and basic heuristics and will influence a performance of the proposed approach. It is assumed that a conditional part of a classifier will describe situations which can be recognized by an agent P_k located in some system node. Objects recognized by an agent are some specific tasks of a program graph. Objects are considered as being "close" to an agent, if a distance to them in a system graph is less or equal then some predefined by a user value of d hops; otherwise, they are considered as being "far" from him.

The following objects are recognised by CS-based agent(see, Fig. 3):

 - task predecessors which data arrives last (bits 1, 2)
 - predecessors with the highest computational cost (bits 3, 4)
 - tasks - not predecessors from the predecessor level of the program graph (bits 5, 6)

- tasks - "brothers" (task from the same level, which have the same predecessor) (bits 7, 8)
- tasks - not "brothers" from the same level (bits 9, 10)
- tasks successors with the highest computational costs (bits 11, 12)
- tasks - not successors from the next level (bits 13, 14)
- waiting time (waiting communication time) for data from predecessors (the difference between the time of receiving data from last and first predecessors) (bits 15, 16)
- waiting time (waiting processing time) for a processor (the difference between the time of beginning processing and the time of receiving last data from predecessor) (bits 17, 18)
- total waiting communication time of tasks located in the same processor (bits 19, 20)
- total waiting processing time in the processor (bits 21, 22)

All these objects can be classified as being all "far", as some of them being "close" or all of them being "close". Some other attributes of a situation in a system graph can be recognised, such as a delay of a given task to be processed on a processor or to a communication channel, a total delay of tasks located in a given node to a processor and to communication channels. These delays can be classified as "acceptable" or "not acceptable" by a comparison with similar characteristics of respectively tasks located in the same system node, or with a situation in neighbouring system nodes.

The following coding of bits in corresponding areas of a classifier is used:
coding 1 (areas between bits 1-14):
00 - all objects are "far"
01 - some objects are "close"
10 - all objects are "close"
11 - not classified situation;
    coding 2 (areas between bits 15-22):
00 - the waiting time is the shortest, comparing with one of the remaining tasks located in the same processor (comparing with neighbour processors)
01 - the waiting time is the same, comparing with one of the remaining tasks located in the same processor (comparing with neighbour processors)
10 - the waiting time is greater than, comparing with one of the remaining tasks located in the same processor (comparing with neighbour processors)
11 - not classified situation.

An action part of a classifier will describe potential actions which can be taken by an agent in a given situation. The actual list contains 16 actions, which are given below (a binary number in parentheses represents coding of each action used in a classifier):

- $a0$: stay in a processor where you are located (0000)
- $a1$: move to a randomly selected processor (0001)
- $a2$: move to a processor, where is located your randomly selected predecessor (0010)

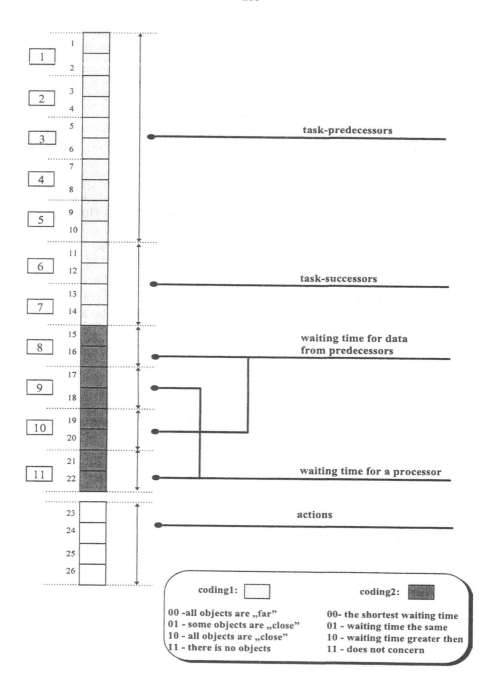

Fig. 3. Structure of a classifier

- $a3$: move to a location of your randomly chosen successor (0011)
- $a4$: move to a processor, where is located a predecessor with the highest computation cost (1000)
- $a5$: move to a processor, where is located a predecessor with a minimal computation cost (1001)
- $a6$: move to a processor, where is located the smallest number of tasks from the same level of the precedence task graph (1010)
- $a7$: move to a processor, where is located the greatest number of tasks from the same level of the precedence task graph (0111)
- $a8$: move to a processor, where is located the greatest number of tasks (1000)
- $a9$: move to a processor, where is located the smallest number of tasks (1001)
- $a10$: move to a processor with the highest computational load (1010)
- $a11$: move to the lowest loaded processor (1011)
- $a12$: move to a processor, where is located the predecessor, which first completed its execution (1100)
- $a13$: move to a processor where are not located "brothers", i.e. tasks from the same level of a program graph, which have the same predecessor (1101)
- $a14$: move to a VIP processor, i.e. to a processor, where is located a predecessor which data arrives last (1110)
- $a15$: move to a processor where is located a predecessor which data arrives first (1111)

Conditional and action parts of a classifier contain, coded with use of binary substrings, and - a don't care symbol # (only conditional part) information concerning the condition and the action, represented by the classifier. For example, a classifier

$$< 1\,0\,\#\#\,0\,1\,\#\#\,\ldots 1\,0\,\#\#\,>:<0\,0\,1\,1>$$

can be interpreted in the following way:
IF (all task predecessor which data arrives last are "close"
AND task predecessors with the highest processing time is not important where (# symbol)
AND task - "brothers" are not important where

. . .

AND total waiting communication time of tasks located in the same processor is greater than the total waiting time in the neighbour processor
AND total waiting processing time in the processor is not important which)
THEN move to the processor of your randomly selected successor.

An initial population of classifiers for each CS is randomly created. In the process of a game the usefulness of classifiers is evaluated. New classifiers are generated with use of GAs and they replace weak classifiers. As the result of the game, the global behavior of the system is expected, resulting in a mapping program tasks to processors, corresponding to optimal or suboptimal scheduling.

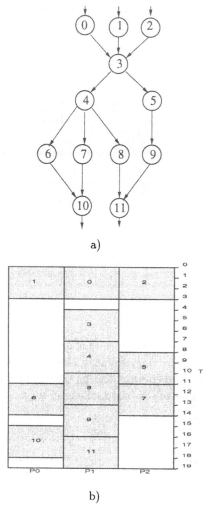

a)

b)

Fig. 4. Precedence task graph (a) and found schedule (b)

5 Experimental Results

In experiments reported in this section it is assumed that a single game is conducted as a sequence of moves (actions) of agents, i.e. at a given moment of time only one agent takes an action. An order of taking decisions by agents is defined by their number in a precedence task graph. A single game is completed in N_p moments of time. A game consists of a number N_G of single games N_g, i.e. $g = 1, 2, ..., G$.

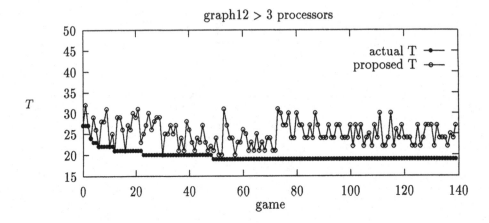

Fig. 5. Performance of CS-based scheduler for the precedence task graph from Fig. 4a

Experiment #1:

In the first experiment a precedence task graph [13] presented in Fig. 4a was scheduled in the 3-processor ring topology (see, Fig. 1a). The precedence task graph has all processing and communication times equal to 3 and 1 respectively. Fig. 5 shows a performance of CS-based scheduler for this precedence task graph.

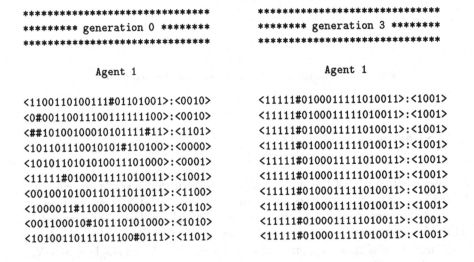

Fig. 6. Population of classifiers of the agent 1 in generation 0 (left), and in generation 3 (right)

One can see that after an initial random mapping tasks of the parallel program into the parallel architecture the scheduler finds an optimal solution with a response time $T = 19$ after about 50 games. Found solution is better than one presented in [13] and obtained applying massively parallel GAs. Fig. 4b describes a solution in a form of a Gantt chart, showing the allocation of each task on processors, and the times when a given task starts and finishes its execution. Fig. 6 shows an initial and final populations of the CS assigned to the task 1. An initial population of classifiers of a size equal to 10 is randomly created. All classifiers have initially the same strength. Twenty games are performed before invoking GA. In the process of a game strengths of classifiers are modified due to a taxation process and rewards obtained by winning classifiers. One can see that for this simple problem a population converges very fast to a homogeneous rule.

Experiment #2:

In the next experiment a precedence task graph [13] with computation and communication times equal to 4 and 1 respectively (see, Fig. 7a) was scheduled into n-processor fully connected system.

To find optimal setting values of a number of parameters which define the CS-based scheduler a set of experiments have been conducted to schedule the task graph into the 5-processor fully connected system. Fig. 8 shows some results of experiments. Each plot in the figure represents the mean of 10 runs of the scheduler. One can see that the minimal response time is found by the scheduler when a population size of classifiers of each CS is greater than 26 (see, Fig. 8a), and a number of games conducted by each CS is at least equal to 20 (Fig. 8b). Fig. 8c shows that about 50% of the best classifiers of each population should be propagated from an old to a new population of each CS, and Fig. 8d shows that winning classifiers should receive a small, equal near to 1, amount of a reward. These values of the CS parameters are used in subsequent experiments.

Fig. 9 shows a performance of the scheduler for an optimal solution and Fig. 10a shows a corresponding Gantt chart. Found solution with $T = 40$ is also better than one presented in [13], where reported $T = 43$.

Table 2 shows a responce time T found for different values of a number of n of a fully connected multiprocessor systems.

Table 2. Response time T for n- fully connected processors

n	2	3	4	5	6	7	8	9	10
T	80	57	46	40	36	33	31	29	29

Found results are the same as ones obtained with use of a standard GA-based scheduler [15].

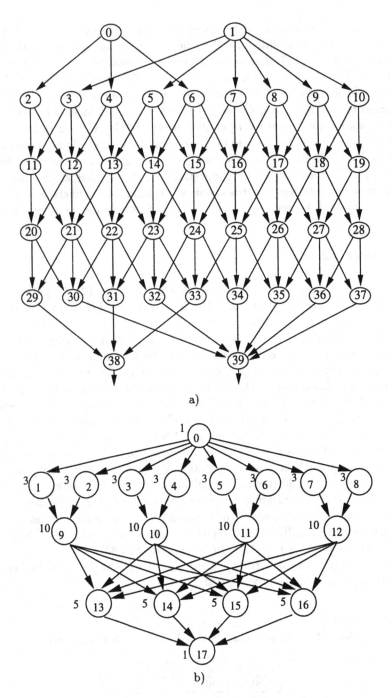

a)

b)

Fig. 7. Precedence task graphs used in experiments

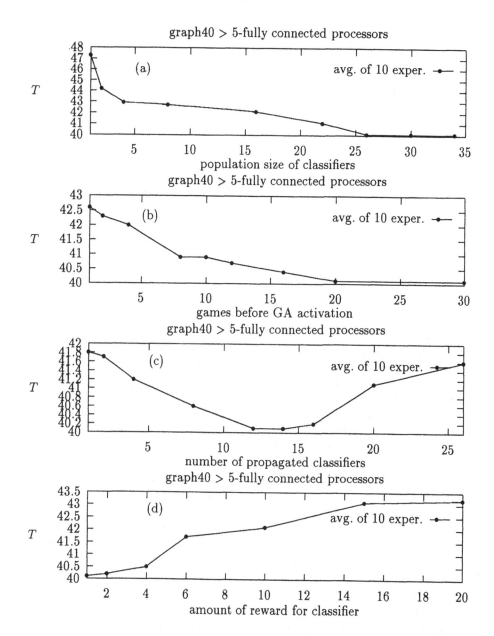

Fig. 8. The average response time T as function of (a) population size of classifiers, (b) number of games before activation of GA, (c) number of propagated classifiers, and (d) value of reward of a winning classifier

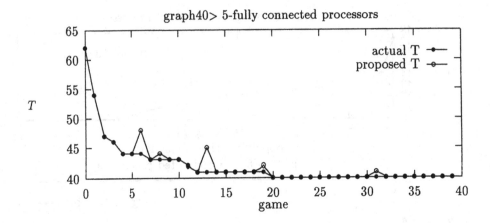

Fig. 9. Performance of CS-based scheduler for the precedence graph from Fig. 7a

Experiment #3:

In this experiment a precedence task graph [8] with all communication times equal to 1, and processing times as shown in Fig. 7b was scheduled in the 8-processor system arranged into one of three popular multiprocessor topologies: ring, cube and grid.

Fig. 11 shows a performance of the scheduler for a ring topology and Fig. 10b shows corresponding Gantt chart with found solution equal to $T = 28$.

Fig. 12 shows frequencies of actions used in the dynamic process of a searching for a solution. We can observe a collective behaviour of some set of active actions (shown in figure), which actively contribute to the process of searching a solution, while the activity of other actions (not shown) serve as a background of the process of searching. This active set of actions is created by the actions *a11*: move to the lowest loaded processor, *a13*: move to a processor where are not located "brothers", i.e. tasks from the same level of a program graph, which have the same predecessor, and *a14*: move to a VIP processor. When the algorithm approaches to the solution an increasing importance of the action *a0*: stay in a processor where you are located, can be observed.

Solutions found for cube and grid topologies correspond to a response time $T = 26$ for both topologies. They are the same as ones found with use of a standard GA-based scheduler.

6   Conclusions

Results concerning ongoing research on development of parallel and distributed algorithms of scheduling tasks of parallel programs in parallel computers have been presented in the paper. Game-theoretic approach to the scheduling with use of genetic algorithm-based learning machines called classifier systems has been

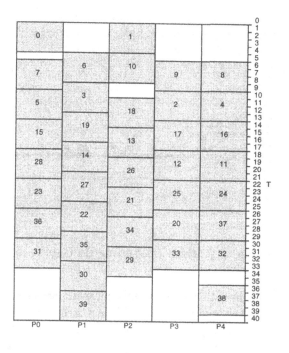

a)

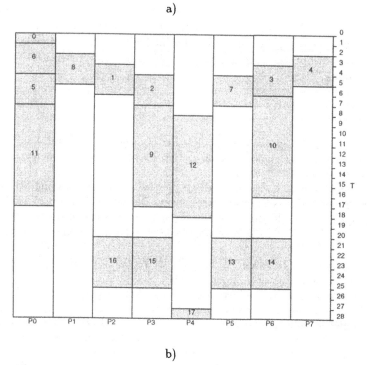

b)

Fig. 10. Gantt charts (a) for the precedence task graph from Fig. 7a scheduled in fully connected 5 processors system and, (b) for the precedence task graph from Fig. 7b scheduled in ring8 topology

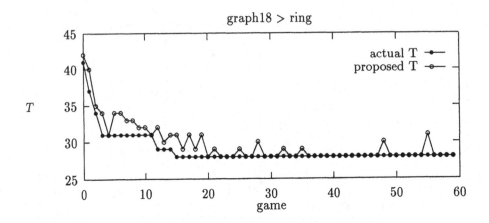

Fig. 11. Performance of CS-based scheduler for the precedence graph from Fig. 7b

proposed. It was shown that such systems have self-organizing futures enabling to emerge a global behaviour in the system. An architecture of classifier system-based scheduler has been proposed. A number of experiments with available in the literature precedence task graphs and different topologies of a multiprocessor system has been conducted. Results of the experimental study of the scheduling algorithm are very promising. In particular they show that the algorithm is able to improve previously known results obtained with use of genetic algorithms. We believe that proposed approach will enable discovering new effective heuristics in scheduling.

References

1. I. Ahmad, (ed.), Special Issue on Resource Management in Parallel and Distributed Systems with Dynamic Scheduling: Dynamic Scheduling, *Concurrency: Practice and Experience*, 7(7), 1995.
2. I. Ahmad and Y. Kwok, A Parallel Approach for Multiprocessing Scheduling, *9th Int. Parallel Processing Symposium*, Santa Barbara, CA, April 25-28, 1995
3. R. Axelrod, The Evolution of Strategies in the Iterated Prisoners' Dilemma. In Davis L. (Ed.). *Genetic Algorithms and Simulated Annealing*. London, Pitman, 1987
4. J. Błażewicz, K.H. Ecker, G. Schmidt, J. Węglarz, *Scheduling in Computer and Manufacturing Systems*, Springer, 1994
5. L. B. Booker, D. E. Goldberg and J. H. Holland, Classifier Systems and Genetic Algorithms, *Artificial Intelligence*, 40, 1989
6. R. Bowden and S. F. Bullington, An Evolutionary Algorithm for Discovering Manufacturing Control Strategies, in *Evolutionary Algorithms in Management Applications*, J. Biethahn and V. Nissen (Eds.), Springer, 1995
7. M. Dorigo and U. Schnepf, Genetic-based Machine Learning and Behavior-based Robotics: a New Synthesis, *IEEE Trans. on Systems, Man, and Cybernetics*, v. 23, 1993

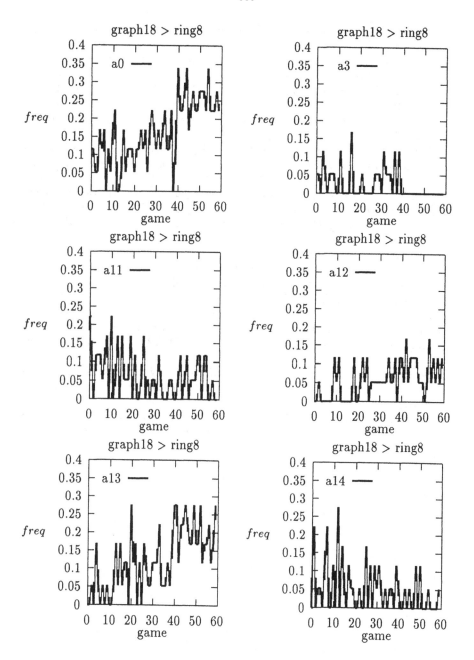

Fig. 12. Frequencies of selected actions

8. H. El-Rewini and T. G. Lewis, "Scheduling Parallel Program Tasks onto Arbitrary Target Machines", *J. of Parallel and Distributed Computing* 9, 138-153, 1990

9. H. El-Rewini, T. G. Lewis, H. H. Ali, *Task Scheduling in Parallel and Distributed Systems*, PTR Prentice Hall, 1994.

10. D. B. Fogel, Evolving Behaviors in the Iterated Prisoner's Dilemma, *Evolutionary Computation*. vol. 1. N 1, 1993

11. D. E. Goldberg *Genetic Algorithms in Search, Optimization and Machine Learning*, Addison-Wesley, Reading, MA, 1989

12. S. Matwin, T. Szapiro and K. Haigh, Genetic Algorithms Approach to a Negotiation Support System, *IEEE Trans. on Systems, Man, and Cybernetics*, v. 21, N1, 1991

13. M. Schwehm, T. Walter, Mapping and Scheduling by Genetic Algorithms, *CONPAR 94 - VAPPVI*, B. Buchberger and J. Volkert (eds.), LNCS 854, Springer, 1994

14. F. Seredynski, Loosely Coupled Distributed Genetic Algorithms, *Parallel Problem Solving from Nature - PPSN III*, Y. Davidor, H. -P. Schwefel and R. Männer (eds.), LNCS 866, Springer, 1994

15. F. Seredynski and P. Frejlak, Genetic Algorithms Implementation of Process Migration Strategies, in *Parallel Computing: Trends and Applications*, G. R. Joubert, D. Trystram, F. J. Peters and D. J. Evans (eds.), Elsevier, 1994.

16. F. Seredynski, P. Cichosz and G. P. Klebus, Learning Classifier Systems in Multi-Agent Environments, First IEE/IEEE Int. Conf. on Genetic Algorithms in Engineering Systems: Innovations and Applications (*GALESIA '95*), Shefield, UK, Sept. 11-14, 1995, IEE 1995.

17. F. Seredynski, Coevolutionary Game Theoretic Multi-Agent Systems, in *Foundations of Intelligent Systems*, Z. W. Ras and M. Michalewicz (eds.), LNAI 1079, Springer, 1996

18. B. Shirazi, A.R. Hurson and K.M. Kavi (eds.), *Scheduling and Load Balancing in Parallel and Distributed Systems*, IEEE Computer Society Press, 1995

Author Index

Lecture Notes in Computer Science

For information about Vols. 1–1251

please contact your bookseller or Springer-Verlag